# TV White Space Communications and Networks

# TV White Space Communications and Networks

Edited by

Robert Stewart
David Crawford
Andrew Stirling

ELSEVIER

WP
WOODHEAD
PUBLISHING
An imprint of Elsevier

Woodhead Publishing is an imprint of Elsevier
The Officers' Mess Business Centre, Royston Road, Duxford, CB22 4QH, United Kingdom
50 Hampshire Street, 5th Floor, Cambridge, MA 02139, United States
The Boulevard, Langford Lane, Kidlington, OX5 1GB, United Kingdom



**Library of Congress Cataloging-in-Publication Data**
A catalog record for this book is available from the Library of Congress

**British Library Cataloguing-in-Publication Data**
A catalogue record for this book is available from the British Library

ISBN: 978-0-08-100611-5



*Publisher:* Mathew Deans
*Acquisition Editor:* Kayla Dos Santos
*Editorial Project Manager:* Lindsay Lawrence
*Production Project Manager:* Surya Narayanan Jayachandran
*Designer:* Vicky Pearson

Typeset by VTeX

# Contents

# Contributors

**Sarbani Banerjee Belur**, Indian Institute of Technology Bombay, Mumbai, India

**F. Rodrigo P. Cavalcanti**, Wireless Telecom Research Group (GTEL), Federal University of Ceará (UFC), Fortaleza, CE, Brazil

**Ciprian Dobre**, Faculty of Automatic Control and Computers, University Politehnica of Bucharest, Bucharest, Romania

**Reijo Ekman**, Turku University of Applied Sciences, Turku, Finland

**Nuno M. Garcia**, University of Beira Interior, Covilhã, Portugal

**Rossitza Goleva**, Faculty of Telecommunications, Technical University of Sofia, Sofia, Bulgaria

**Farzad Hessar**, University of Washington, Seattle, WA, United States

**Marko Höyhtyä**, VTT Technical Research Centre of Finland Ltd, Oulu, Finland

**Jussi Kahtava**, Allied Spectrum Associates Ltd, Fleet, United Kingdom

**Juha Kalliovaara**, Technology Research Center, University of Turku, Turku, Finland

**Abhay Karandikar**, Indian Institute of Technology Bombay, Mumbai, India

**Meghna Khaturia**, Indian Institute of Technology Bombay, Mumbai, India

**Arto Kivinen**, Turku University of Applied Sciences, Turku, Finland

**Heikki Kokkinen**, Fairspectrum, H2020 Coherent, Espoo, Finland

**George Mastorakis**, Technological Educational Institute of Crete, Agios Nikolaos, Crete, Greece

**Constandinos X. Mavromoustakis**, University of Nicosia, Nicosia, Cyprus

**Jordi Mongay Batalla**, National Institute of Telecommunications, Warsaw, Poland

**Miia Mustonen**, VTT Technical Research Centre of Finland Ltd, Oulu, Finland

**Jarkko Paavola**, Turku University of Applied Sciences, Turku, Finland

**Sumit Roy**, University of Washington, Seattle, WA, United States

**Carlos F.M. e Silva**, Wireless Telecom Research Group (GTEL), Federal University of Ceará (UFC), Fortaleza, CE, Brazil

**Andrew Stirling**, University of Strathclyde, Glasgow, United Kingdom

**Pekka Talmola**, Nokia, Espoo, Finland

**Xuhang Ying**, University of Washington, Seattle, WA, United States

# Preface

*TV white space communications and networks* seems a dry title for a new approach to accessing spectrum that promises to revolutionise wireless connectivity. The combination of sub-1 GHz TV band frequencies and licence-exempt, dynamic spectrum access has been used to radically extend broadband coverage in rural and urban areas.

The focus of wireless networking innovation has been on increasing the capacity to cope with the rise in the use of smart mobile devices and the ever more powerful cloud-based services they link with. However, lack of coverage is also a problem, especially as policymakers and leading industry players are seeking to provide the economic and social benefits of getting online to as much of the world's population as possible.

By harnessing unused UHF spectrum, TV white space communications technology can help fill un-economic coverage gaps using established technologies and business models. Access to this fragmented spectrum, intimately entwined with live television broadcast services, requires a new approach to regulation termed Dynamic Spectrum Access (DSA). This new mechanism enables the safe coexistence between the established licensed services and new licence-exempt users in a way that could ultimately be used to allow flexible and rapid access to any part of the spectrum.

The book provides an introduction to and exploration of the key themes and issues associated with TV white space communications. It aims to provide the reader with an overview of the available technology and regulatory trends, and offers greater depth in TVWS fundamentals to those who want to know more. In this volume:

- Illuminating contributions from leading experts in this field have been brought together.
- TV white spaces are defined and explained, illustrating how unused TV band spectrum can be applied to extending affordable Internet access.
- The policy developments which have led to this new opportunity are summarised.

- Geolocation databases are introduced; this is a new fundamental spectrum management tool which enables more flexible, granular spectrum access to support wireless innovation.

In addition to the mechanics of TV white space communications from regulatory and industry perspectives, the book explores some key application areas, including affordable rural broadband and machine-to-machine communications (such as Internet of Things applications).

Terrestrial TV transmission bands are gradually being rolled back as governments across the world seek additional capacity for mobile broadband. Whilst this would seem to threaten the eclipse of TV white space communications, this book illustrates how these frequencies can deliver additional capacity (and coverage) for connectivity without needing to wait for the painfully slow and expensive process of clearing the broadcasting bands. Given the political sensitivities around terrestrial broadcasting and the need for international agreement, it is likely that broadcast bands will be with us for many years to come, so TV white spaces technology will have ample opportunity to demonstrate its value. The DSA, which is the key to opening up this unused spectrum, also lays the foundation for more flexible and efficient management of a much broader range of frequencies in the future. The DSA will enable regulators to adapt spectrum supply more quickly to rapidly evolving market needs.

(The authors are all connected with the University of Strathclyde's *Centre for White Space Communications*, a centre of excellence for research into technology for dynamic spectrum access and its applications to enhancing connectivity worldwide.)

**Robert Stewart**
**David Crawford**
**Andrew Stirling**

# About the Editors

**Andrew Stirling** has played a key role in advocating the opening up of TV white spaces in the UK and other parts of Europe.

With a degree in Physics from Imperial College, London, Andrew joined the BBC's Research & Development Department to work on digital innovations for TV and Radio. Andrew later worked on digital switchover policy, at the United Kingdom's Ofcom. His consultancy, Larkhill, has advised BBC, Dell, Microsoft and Intel particularly on spectrum policy and connectivity.

From 2011 to 2012, he directed the landmark Cambridge TV White Spaces Trial, attracting nearly 20 leading players in Communications, from start-ups to multinationals.

Since the enactment of UK regulations in 2016, he has supported the first commercial broadband roll-out making use of TV white spaces technology, delivering rural broadband to poorly connected communities on the Isle of Arran, Scotland.

Andrew is Chair of the *Centre for White Space Communications* and Honorary Lecturer at the University of Strathclyde.

Andrew has written papers, articles, reports and regulatory submissions on the subject of Dynamic Spectrum Access (especially on its initial application to the TV white spaces) and he has also represented the Dynamic Spectrum Alliance in international conferences at the ITU and in key European regulatory working groups.

Currently the MathWorks Professor of Signal Processing at the University of Strathclyde, since August 2014 **Robert Stewart** has been Chair and Head of the Department of Electronic and Electrical Engineering, where there are more than 55 academic staff and almost 300 researchers. His interests in teaching, knowledge exchange and research over the last 20 years have focused around signal processing and communications. In recent years Prof. Stewart has been working on digital communications and software-defined radio, with a specific interest in radio standards such as LTE, 802.11 and most recently on wireless white space radio utilising TV spectrum frequencies. From 2006–2012 he was the Xilinx Professor of DSP and Digital Logic at Strathclyde, and from 1997 he has been a visiting Professor at the University of California, Los Angeles (UCLA), Extension School. In 2004 Prof. Stewart was a founder of the technology company Steepest Ascent Ltd., based in Glasgow with an office in Los Angeles. The company was acquired by MathWorks in 2013, which now has an engineering development office in Glasgow near the University.

**David Crawford** runs the Centre for White Space Communications, an industry-focused R&D centre set up by the University of Strathclyde in March 2011. The Centre works with industry partners to investigate techniques and applications for the efficient utilisation of white space radio spectrum. Dr. Crawford is a Chartered Engineer with more than fifteen years' senior engineering and management experience in the consumer electronics industry. He has a PhD in Electronic Engineering and an MBA, both from the University of Strathclyde.

Chapter 1

# TV White Space Developments in the UK

**Andrew Stirling**
*University of Strathclyde, Glasgow, United Kingdom*

## 1.1 INTRODUCTION

The United Kingdom was not the first country to consider regulations for TV white spaces (TVWS). However, following its detailed and painstaking implementation the UK regulatory model has been influential as other regulators around the world seek a solution that fits their needs.

From the introduction of mobile services in the 1990s, the appetite for spectrum to support the growth of wireless services has developed quickly. The perceived need to add data capacity to mobile networks fuelled the interest of operators, who bid a total of £22 billion in the auction for 3G licences in the United Kingdom in 2001. This unexpected windfall, in turn, increased Government interest in making more spectrum available, and led to a review of spectrum management (by Professor Martin Cave) and to a mandate (and budget) for the new UK regulator Ofcom to conduct research on how to manage spectrum more efficiently.

It was against this backdrop that digital TV switchover was kicked off with the aim of freeing spectrum in order to exploit the greater efficiency of digital broadcasting technology. At the same time, Ofcom was open to gaining additional economic benefits from the application of new technology to tap into unused spectrum. Thus, the first glimmerings of UK regulatory interest in TV white spaces (TVWS) began.

## 1.2 WHAT ARE TV WHITE SPACES?

TV white spaces are essentially unused spectrum in the frequency bands allocated for terrestrial television broadcasting. As with holes in materials, they need to be defined in terms of the absence of something rather than being entities in their own right.

**TV White Space Communications and Networks. http://dx.doi.org/10.1016/B978-0-08-100611-5.00002-3**

In the United Kingdom, the frequencies used for terrestrial TV broadcasting range from 470 to 790 MHz (which accounts for a significant fraction of UHF Bands 4 and 5).

The white spaces can be considered in terms of time, place and frequency. In any given location:

- Particular frequencies may be available if the nearest TV transmitters are not making use of them;
- The availability of frequencies might be restricted in time, for example if PMSE applications (e.g. wireless microphones) are making use of the frequencies at the location in question.

The nature of TV broadcasting means that the frequencies available in one location are often different to those available in another location. Indeed, the spectrum allocated to broadcasting is also referred to as the *interleaved spectrum* because of the 'patchwork quilt' of frequencies used by transmitter stations to provide coverage across the nation and/or across a region for each television network. This fragmentation has complicated the allocation of the spectrum to potential uses. Typically, licences are granted for large areas because this enables efficiency for the users and for the administrations and regulators. The smaller the area to be licensed, the smaller the potential value compared to the transaction costs involved in licensing and the costs in provisioning a network to cover it.

## 1.3 WHAT DOES AVAILABLE MEAN?

Availability of a frequency does not mean empty or unoccupied but merely that whatever signal might be carried on those frequencies is not needed or is not protected by regulation. All frequencies have a level of noise and interference. Background noise arriving from outer space provides the ultimate floor, whilst man-made radio sources raise this floor by varying degrees, depending on the time and place.

In the TV band, signals from distant transmitters are the most likely reason for 'noisy' but available UHF frequencies, sometimes termed 'grey spaces'. Considering an example from the south-east coast of England, a viewer may be able to receive a range of terrestrial services from the BBC and other UK broadcasters, as the service planners intended. However, it might also be technically possible for the same viewer to receive signals from French broadcasters on other UHF frequencies. These latter frequencies would be regarded as white spaces by the UK regulator Ofcom since it has no duty to protect the reception of French TV services by viewers based in the United Kingdom.

## 1.4 WHY TV BAND SPECTRUM IS VALUED

The UHF frequencies offer enhanced propagation compared with the much higher frequencies used for 3G and Wi-Fi networks in the 2 GHz and 5 GHz bands. This translates into:

- Coverage extending further from the base-station/access point, enabling sparser and more cost-effective infrastructure for a given transmission power;
- An ability to serve locations which are not directly in sight a base station or which are sheltered by woodland;
- Improved indoor coverage.

Another factor in favour of the TV white spaces is that TV broadcasting occupies similar frequency ranges around the world, providing potential for device volumes that would help to make the technology more affordable.

## 1.5 ORIGINS OF TVWS

TV white spaces arise for two reasons:

1. The way that terrestrial television services are planned: high tower, high power transmitter networks are used to achieve cost-effective coverage with only relatively simple and inexpensive hardware installations being required at the receiving end. Because of the range of the television broadcast signals, TV broadcast planning is an international exercise. To provide each European country with the capacity it needs for regional and national networks, the 32 UHF frequencies currently available for broadcasting (extending from 470 to 790 MHz) need to be reused intensively across the continent. With the traditional multi-frequency approach to Digital Terrestrial Television (DTT) broadcasting, neighbouring transmitters need to use different frequencies in order to avoid mutual interference and the resulting destruction of coverage. Under this model, there is a minimum separation required before a frequency can be reused at another transmitter site. Thus, the established terrestrial TV distribution model naturally leaves many gaps in frequency use, which we refer to as the TV white spaces.
2. It is not economically feasible to serve certain remote rural locations with terrestrial broadcasting. This is a more significant factor in countries such as the United States, where broadcasters are more commercial and tend to be city-based. The UK/European broadcast licensing model has enabled regional and national networks, with coverage across the licensed region or nation in rural and in urban areas. However, because the United Kingdom is relatively highly urbanised, around 90% of households can be covered with transmitters providing coverage reaching only just over half of the landmass

of the country. The TV spectrum is left unused outside of these areas for economic reasons, and therefore white spaces arise.

### 1.5.1 A Short History of Terrestrial TV in the United Kingdom

Terrestrial television services in the UK commenced as early as 1929, following the pioneering work of John Logie Baird and others, just a few years after the start of public radio broadcasting. The first television service provided by the British Broadcasting Corporation made use of spectrum in the VHF (Band III). By the 1960s, advances in television led to the introduction of further higher definition services in colour. These later services used spectrum which had been cleared in UHF (Bands 4 and 5) under international agreements (facilitated by the ITU).

The initial transmitter sites, such as Crystal Palace in South London, were chosen with the aim of serving a large number of households from a single site. Over time, other sites which served smaller and smaller pockets of population were added, eventually reaching 1154 sites in total across the United Kingdom.

At each transmitter site, only a handful of the available 48 UHF channels (frequencies) were used to carry the relatively modest number of TV networks that were broadcast to UK audiences. Up to 1998, when the first digital terrestrial TV (DTT) services were introduced, there were a maximum of 5 (analogue) TV networks. The 6 DTT multiplex networks, which were simulcast with the analogue services, brought the total number of networks to 11 (with each DTT multiplex network carrying several digital TV channels), so that around each transmitter site there were at least 37 (i.e. 48-11) frequencies left unused.

Switching off the analogue services freed up to 5 frequencies. However, the introduction of two temporary multiplexes to support the clearance of broadcast services from the 700 MHz band and a local TV multiplex in some areas brought the total number of occupied UHF channels to up to 9 at each transmitter site.

## 1.6 THE UK ROAD TO TV WHITE SPACES REGULATION

### 1.6.1 A Brief Overview of UK Spectrum Regulation

Spectrum regulation in the United Kingdom has grown from being a pure administrative matter, referred to as a 'command and control' approach [1] to being more open and responsive to market demands.

Traditionally, spectrum regulation had been an 'under the hood' activity, safely left in the hands of engineers and technocrats who had a better grasp of the practicalities of wireless technology. It had been the applications, particularly broadcasting and telecommunications, which were the focus of policy makers. Broadcast policy would tend to determine the scope and priorities for the other communication sectors. Spectrum regulators were left to ensure that broadcast needs were met, and that access to the remaining capacity was efficiently distributed and coordinated with the other services.

The introduction of digital terrestrial TV broadcasting was supported by policy makers because it was clear that the terrestrial television service was in danger of falling behind the multi-channel offerings of satellite and cable operators. Digital broadcast technologies offered an increase in spectrum efficiency. For example, 10 television channels could now be carried in the same bandwidth as a single analogue TV channel.

Broadcasters were keen to use the new technology to multiply their own distribution capacity, retaining the same number of UHF channels (48) as before.

Policy makers, on the other hand, saw other possible gains:

1. Creating opportunities for new broadcasters leads to increased choice and innovation which benefits consumers and viewers;
2. Releasing spectrum for new applications feeds the market's rapidly growing thirst for mobile data capacity.

Policy makers increased the effective capacity for broadcasters by applying the new digital broadcast technology; they were able to release a total of 112 MHz for new applications (with the growing needs of mobile broadband services very much in mind). This released spectrum was referred to as the digital dividend. TV white spaces offered the bonus of additional capacity to the digital dividend, but without the need to reduce digital broadcasting capacity.

### 1.6.2 Ofcom's Central Role in UK Spectrum Management

The UK communications regulator is Ofcom, which was established by the UK Communications Act (2003). Its duties encompass broadcast and telecommunications, as well as spectrum management (for civil use), and it has taken a mainly economics-oriented approach to spectrum management by welcoming innovation towards more efficient spectrum use. Recognising the growing appetite for mobile data capacity, Ofcom has been actively seeking spectrum that could be made available for new applications. This is the background to the United Kingdom's openness to proposals to make use of TV white spaces.

Ofcom's approach to creating capacity for new applications is principally to clear out older technologies (as it did with analogue television broadcasting), but it has also come to appreciate that more intensive sharing of bands is a way to create new capacity, i.e. allowing new applications to fit alongside established ones. This approach avoids the cost and delay of clearing spectrum, but has created the need for a new regulatory approach, backed by technologies such as Dynamic Spectrum Access, which is the technical basis for enabling access to the TV White Spaces.

### 1.6.3 Traditional Spectrum Management Uses Dedicated Allocations and Licensing

Spectrum access is normally provided through licences which are granted by Ofcom. Licensing enables Ofcom to manage the use of spectrum so that users do not disrupt each other's wireless services and applications. This ensures that licensees are protected against harmful interference and also secures guaranteed access to market. Revenues from licensing then support Ofcom's work in policy development and enforcement.

Ofcom's licensing sits within an international framework of spectrum management to ensure that use of spectrum in one country does not impair services in adjacent countries.

Traditionally, spectrum has been allocated to particular applications on a band-by-band basis, ultimately through the World Radio Conferences held every four years by the International Telecommunications Union (ITU). Although there is a strategy to move away from dedicating particular bands to particular applications under the European Wireless Access Policy for Electronic Communications Services (WAPECS) approach, the band-by-band model still applies broadly because it simplifies the technical requirements for coexistence. Whilst allocation happens on an international level, it is the job of the individual national administrations (and their spectrum regulatory bodies/departments) to actually assign licences to the ultimate users of spectrum in each nation's domain. National regulators also need to enforce the terms of licences, such that each nation's international commitments are fully met and to prevent harmful interference to services in neighbouring states.

In certain bands—such as those allocated for broadcasting and mobile communications—there is greater demand for spectrum than there is capacity available to meet it. In these cases, Ofcom awards spectrum access (licences) through a competitive process:

- Beauty parades where bidders submit proposals, e.g. for a new television service, which are judged against a range of criteria;
- Auctions where suitably qualified bidders compete in terms of the sum of money they are willing to pay for the licence and for the rights and responsibilities it brings.

The ultimate winner(s) of the competition will then be subject to the technical conditions specified in the licence, for example

- permitted locations of transmitter installations
- antenna heights
- effective radiated power.

A licence provides permission for an operator to use the spectrum in a carefully defined way, but ownership of the band is retained by the nation and the regulator is normally free to assign licences to other users in the same band provided that no harmful interference is caused.

### 1.6.4 Licence Exemption – Assumed Access Permission

As well as licensed use of spectrum, it is possible to make use of certain bands on a licence-exempt basis. UK regulators prefer to use the term *licence exemption* rather than *unlicensed* since the latter encompasses illegal use of spectrum.

This 'general authorisation' is subject to technical conditions which are laid down in the enabling legislation. With licence exemption, the user can presume that they are authorised to use the spectrum provided that the equipment they are using complies with the relevant standard(s) such as those laid down by the European Telecommunication Standards Institute (ETSI) for European countries.

The equipment manufacturer picks up the task of ensuring that the technical requirements of the licence exemption are met with little or no dependence on the end users' skill or competence.

An essential feature of licence-exempt spectrum use is that the regulatory authorisation is not exclusive. Furthermore, there are no fees to be paid for access to spectrum. Thus, licence exemption opens up spectrum access to a much wider range user types than is practical with licensing, from individuals at home to major corporations deploying wireless networks across a wide area.

Licence-exempt technology is normally preferred by manufacturers for inclusion in devices since it is often less expensive in terms of components and intellectual property costs. Harmonised licence-exempt bands such as 2.4 GHz and 5 GHz enable high volumes which enable Wi-Fi and Bluetooth technologies to be affordable. Given the worldwide harmonisation of terrestrial TV broadcasting bands, TV white spaces technology could enjoy the same economies of scale, helping it to achieve proliferation in smart devices.

The end users are the beneficiaries of licence exemption; they are able to deploy equipment to provide their own coverage whether in a building, on a campus or on a farm, for example.

### 1.6.5 International Harmonisation of Spectrum Management

International coordination is facilitated by the ITU. Its work, which draws on experts from member states across the world, has been vital to enabling the advances in wireless communications over many decades which are enjoyed internationally. Gaining ITU agreements on the broad allocations of spectrum for television and telecommunications is a slow and expensive process, but has been key to enabling sufficiently large markets to support further development of wireless technology.

Now, however, the pace of innovation is such that the time required for ITU processes has become a hindrance to progress. Increasingly, countries are using their own flexibility to accommodate new technologies within broader international frameworks. TV white spaces are an example of how a new technology can be included within an established framework of the international digital broadcasting frequency plan without the need for specific recognition by the ITU. Ofcom was able to develop its enabling regulations for TVWS to comply with what had already been agreed to internationally, within ITU Region 1, in terms of the United Kingdom's responsibilities to its neighbours.

### 1.6.6 UK TV White Spaces – The Enabling Regulations

Work on TV white spaces began as part of Ofcom's review of the potential of a 'digital dividend' from the switchover from analogue to digital television. The analogue and digital broadcast systems were in operation together (simulcast) from 1998 to 2013; the switchover process began its sweep across the UK in 2006. This presented the prospect of releasing five frequencies at each transmitter site and added to the available white spaces capacity.

Whilst the main thrust of the digital dividend effort was towards freeing up UHF spectrum for new applications and keeping in mind enhanced mobile data provision, Ofcom was prepared to take a broader view of capacity for new applications. Leading companies from the technology sector pointed out the broadband coverage and other innovation benefits that could be gained from having a range of UHF frequencies available on a licence-exempt basis. These frequencies, which were much lower than those previously applied to wireless broadband access, would help extend broadband into the many, often rural, areas which were not yet adequately served.

### 1.6.7 Basis of Protection for the Licensed Services

Ofcom made it clear in its 2007 digital dividend Review Consultation that it would consider enabling access to the TV white spaces on the condition that there should not be any harmful interference to the established licensed services, TV broadcasting and PMSE applications.

### 1.6.8 Licence Exemption – The Default Access Model for TV White Spaces

Since the broadcasters and PMSE users do not own the band they use—they only have permission to use certain frequencies in certain locations—Ofcom determined that it needed to manage the assignment of the remaining, unused, spectrum. Given the fragmented nature of this TVWS spectrum, and industry requests for innovation flexibility, Ofcom pursued a licence-exempt model for access.

Ofcom convened a working group with industry representatives to work on potential solutions to ensure that licence-exempt access to the white spaces could coexist with the licensed services.

The new white spaces applications would not be permitted to cause harmful interference to the established services nor would white space users be able to claim rights to protection from other users. Thus, white space access would be opportunistic: capacity would be available depending on what was left over from the licensed services, and there would be power constraints on white space devices (WSDs) to protect those services. White space capacity could not be guaranteed since the licensed services have higher priority. PMSE applications (such as wireless microphone links), particularly around theatres and studios, are heavy users of white space channels.

In seeking a solution to coexistence with licensed services in UHF, Ofcom had to consider how devices would know:

which channels they would be allowed to use in their location and
what effective radiated power they could emit on each of the available channels.

## 1.6.9 Determining Which Channels Would Be Available

It is important that a WSD avoid frequencies which are in use by the licensed services in the device's location.

Ofcom considered three principal models for determining available frequencies, following the lead established by the United States:

- Spectrum sensing
- Beacons
- Geolocation databases.

### Spectrum Sensing

Sometimes termed 'cognitive radio', this method of determining available spectrum was favoured by the industry because of its simplicity and lower cost of implementation. Technology had been developed (including energy-based and more sensitive detection methods, e.g. cyclo-stationary). The state of the art was capable of detecting a TV signal at a thousandth of the power level actually required by TV receivers to satisfactorily decode the services.

However, the broadcast industry was concerned that under certain conditions even this sensitivity would be insufficient, for example in a case where the

WSD is shielded from the TV transmitter emissions but not from a DTT receiver whose reception it could disrupt.

The sensing requirements which resulted from broadcaster concerns were of such a level that they would be difficult or expensive to implement, and would also be likely to yield a high rate of 'false positives'. In other words, the costs of the devices would be high and the efficiency of spectrum use would be low.

Sensing by itself would also be unable to take into account different vulnerability levels of the licensed services, which vary by location across the United Kingdom. For example, there are relatively few UHF wireless microphones needing protection in the Cairngorm Mountains in the Highlands of Scotland.

### Beacons

Another way for WSDs to determine which frequencies are available is for a pilot or marker signal to be provided. These transmissions can be provided in a given area to enable WSDs to know which operating frequencies are safe.

Ensuring that there is beacon coverage over the United Kingdom would be an expensive undertaking. Ongoing operating costs would be another issue, and the business model is far from clear.

There would be technical issues as well.

- To minimise beacon coverage costs, it would be desirable to have a large area served by each beacon transmitter. However, this would conflict with spectrum efficiency since frequency availability might vary across the beacon's area of service.
- Devices at the edge of the beacon area might transmit into a neighbouring area where frequency availability differs.
- Beacon signals would themselves occupy spectrum without delivering any direct value.

### Geolocation Database

Geolocation databases make use of the position of a WSD in order to provide it with safe frequencies to use. The frequencies available in each area (the United Kingdom uses 100 m square pixels in its geolocation databases) could be calculated by the data, given the position of DTT transmitters (and studios/theatres, etc., in the case of PMSE users).

Under this model, WSDs would obtain a list of available frequencies from an approved geolocation database. The list's validity would hold

- within 15 minutes of the time the frequency list was provided;
- within a 50 m radius of the location given by the WSD when the list was requested.

Outside of these limits, the WSD must request a new frequency list and report its position again.

Geolocation databases provide certainty for the established licensed services since it is then clear for each possible location which frequencies are available for WSDs. They also provide clarity for WSD users who can see how much capacity is available.

There are no guarantees regarding the spectrum capacity that will be available at a given time, however, because PMSE users can make last-minute bookings which reduce the capacity available for other WSDs. (PMSE devices are themselves WSDs, albeit operated under user licences.) DTT usage of spectrum tends to change much less quickly.

There is also no certainty with regard to how 'clean' the available frequencies are. Just because a frequency range is not protected in a given location does not mean that the interference and noise level will be low. There may be overspill from areas where coverage is protected into areas where the wanted signal is considered to be below the minimum level deemed sufficient for coverage to be declared and protected.

Databases are now central to managing many aspects of everyday life including money, airlines seats and a whole range of assets and transactions. They enable scaling efficiencies by automating the management of records, increasingly via devices/systems which host services for end users such as flight ticketing, etc.

For the management of spectrum, databases offer a number of advantages:

- They provide a convenient tool for hosting data on spectrum assignments and associated rules, which can then be scaled to much finer levels of granularity than would be practical with manual assignment processes and record keeping.
- With international spectrum management being fragmented across the countries of Europe, it is easier to accommodate the differences in a database, allowing harmonised devices to be produced for an international market.
- The database contents can be changed relatively easily and quickly, allowing regulators to correct for any issues that might arise either locally or across the whole of a nation or region.

## Ofcom Chose the Geolocation Database Approach

After extensive consultation, Ofcom proposed geolocation databases to enable licence-exempt access to the TV white spaces because of the certainty and clarity of the protection it affords to existing licensees. Ofcom could also see the benefits of applying geolocation databases to a much wider range of frequencies than just TV white spaces.

In parallel with developing the database mechanics, Ofcom contributed to an ETSI working group to determine the technical standard with which the WSDs would need to comply. The reference for this EU harmonised standard is EN 301 598. Among other things, the standard lays down requirements for the emission mask for the device; 5 classes of device are allowed according to how stringent a mask the device can comply with. Class 1 devices have the cleanest mask and benefit from the greatest access to white spaces spectrum. At the lower end of the scale, Class 5 devices are more restricted in the frequencies and power levels that they are allowed use.

Ofcom also developed mechanisms for helping itself to locate and deal with any potentially harmful interference such as using database audit access and rapid notification of instructions to cease using a frequency.

### Determining the Transmitted Power Limits for White Space Devices

In order to determine what power levels WSDs can safely emit, Ofcom had to create a reception model for the licensed services, making assumptions that would be representative of real life.

A key parameter to be considered is the effective sensitivity of the receiver of the licensed service (digital terrestrial television) since the receiver would need a certain margin for the desired signal above the background (i.e. the combined 'noise and interference' floor).

The analogy of a boat can be used here. If we imagine the receiver margin to be equivalent to the depth of water above the river or sea bed, then unless the depth is at least as great as the submerged section of the boat, it will not be able to sail freely.

If DTT is the wanted (protected) signal, then the unwanted signals and noise level would now include transmissions arriving at the receiver from WSDs and from other emitters.

Ofcom determined how low these received emissions from WSDs would have to be to ensure a sufficient margin for the receiver. It then worked backwards to determine what the limit of radiated power from the WSD would need to be. Ofcom also needed to make assumptions about the loss of signal strength along the signal's path between the WSD and the DTT receiver antenna, also referred to as the 'coupling gain'. Data provided by the WSD, such as antenna height and modulation type, have an impact on what power limits are notified to the device by the database.

Ofcom could have chosen the lowest possible signal level for the DTT receiver to function as the level to be protected, but instead it chose the level used by planners to determine which areas have reliable coverage for a DTT signal. This allowed WSDs more headroom for their emissions and hence wider coverage potential.

A further critical issue was Ofcom's choice of the rooftop antenna as the TV reception mode to be protected, again in line with the way DTT services are planned. If the more sensitive 'set-top/portable' antenna mode had needed protection then, again, WSD emissions would have been more severely limited in power largely because the distance between the WSD and the TV antenna would be much smaller and therefore a given WSD emission level would have more impact.

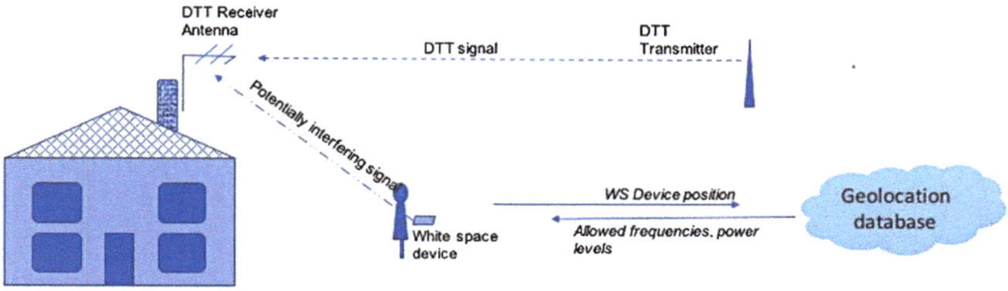


To determine the allowed WSD emission power, provided via the database the regulator makes cautious assumptions about the signal path losses between the WSD and DTT receiving antenna.

In its modelling, Ofcom assumed that WSDs could be personal and portable, much as mobile phones are. The present reality is that the current generation of WSDs are too bulky and power hungry to be pocketable; instead, they are typically installed on rooftops or gable ends and close to their antennas.

Housing density was another factor to be considered because a neighbour's use of a WSD might impact another's DTT reception.

In practice, there are a number of factors that would mitigate the impact that WSDs have on DTT reception:

- Current WSDs are normally professionally installed with fixed antennas, reducing the potential for harmful interference to DTT reception by far more than Ofcom assumed in its model;
- Personal portable WSDs, expected in the future, are likely to have their emission attenuated by antenna constraints and the way the end users carry or wear them, much as mobile phones (and wireless microphones) are;
- Because of the sensitivity of early DTT services to emissions from faulty vehicle ignition systems, etc., rooftop antennas have often been installed in such a way as to obscure their view of nearby roads;
- Most TV viewing occurs in the evening when there are fewer people, and therefore fewer WSDs being taken around.

## Geolocation Databases

Under the UK framework, WSDs must obtain a list of safe frequencies from a validated and recognised geolocation database. The database's role is to host the frequencies and power levels which are calculated by Ofcom and transferred to each database service provider. The database service providers also need to take the set of PMSE assignments and to calculate the remaining available frequencies and power levels to ensure that both DTT and PMSE services are adequately protected.

A number of geolocation database providers, including Fairspectrum, Google, Microsoft, Nominet and Sony, have had their service qualified through an exhaustive test programme operated by Ofcom. The qualified providers are listed in the UK licence exemption regulation for WSDs. However, only two providers are currently listed as active on Ofcom's database discovery web page, Nominet and Fairspectrum.

More details on geolocation databases are provided by H. Kokkinen of Fairspectrum (see Chapter 2).

## Manually Configurable Device Licences – An Interim Measure

Ofcom's original intention (and the industry's desire) was to enable licence-exempt access to the TV white spaces. This would be achieved by ensuring that the regulatory framework and device standards inherently protect the licensed services without relying on the end users in any way.

However, at the time when the UK TVWS licence exemption regulations came into force (January 2016) there was no equipment available to meet Ofcom's requirements in full. In particular, the available WSDs lacked the facility to locate their position (geolocation). Devices were already on the market in the United States, where it was permitted for their professional installers to enter the device position manually during the system installation. However, Ofcom and the licensees were not happy to have manual position entry for geolocation; they insisted that it should be automatic and integrated.

To enable use of the available manually configured TVWS radios, Ofcom decided to introduce a licence. The device would have to meet the ETSI requirements that frequencies should be obtained from a geolocation database, but it would be the licensee (normally a service operator) who would be responsible for ensuring the correct installation and entry of the device's position.

Ofcom envisaged that the Manually Configurable Device (MCD) licence, would be an interim measure with a three-year validity. After this, it was hoped that devices with automatic geolocation would become widely available. To take advantage of the licence, operators must satisfy Ofcom of their competence and pay (currently) £1500 per year. There is no limit to the number of sites on which they can deploy the equipment.

## Regulatory Requirements on Licence-Exempt Devices

Licence exemption puts the onus on manufacturers of equipment to satisfy the technical standards and rules laid down under the regulator's terms. The end users cannot be presumed to have technical knowledge about how the devices they have purchased will behave. In Europe (including the United Kingdom), they can instead rely on 'CE' marking, which is the indication that a product meets all relevant EU requirements (including spectrum access for radio transmitter and receiver products).

White space device manufacturers that wish to market their products in the United Kingdom need to take account of the compliance requirements which apply across the EU. This self-certification regime gives manufacturers the freedom to choose how they will prove compliance with the relevant regulations, rather than being required to submit devices to a single national or regional body.

The WSD manufacturers need to consider which EU directives apply to their device (including those applying to electrical safety and immunity to electromagnetic interference). The radio requirements for WSDs are laid down in an ETSI harmonised standard (EN 301 598) which is recognised across Europe. However, the geolocation databases that WSDs would need to interface with are only open to work with white space devices located in the United Kingdom at this point.

## The United Kingdom's Regulatory Journey Began in 2005/2006

Around 2005, Ofcom started to work on defining a *digital dividend*, a bundle of frequencies to be freed up following the switchover from analogue to digital television. At the industry's request, Ofcom also started to consider the possibility of using the spaces left between the TV network transmissions. Although the extraction of the digital dividend resulted in the spectrum retained for TV transmissions being reduced from 48 to 32 channels, the closure of the analogue transmissions helped to mitigate the loss of white spaces.

There was an established precedent for white spaces use: broadcasters, theatres and other venues had found these spare pockets of spectrum useful in supporting the 'back office' for productions. Referred to as Programme Making and Special Events (PMSE), the applications range from wireless microphones and wireless cameras to in-ear monitoring devices. At major productions there can be tens of wireless microphones in use, mopping up many of the white space channels in the vicinity of the venue. In the United Kingdom these applications have been managed for many years using a light licensing approach with registrations administered by Ofcom.

A major difference with the proposed new TVWS applications was that the devices could be available to a mass market with licence-exempt access to the spectrum.

Ofcom considered the initial industry proposals and decided to commit to opening up access to the TV white spaces provided there was no negative impact on the existing licensed services, DTT and PMSE.

Ofcom first had to consider who should make the decisions regarding TV white spaces, Ofcom itself or the broadcasters. Since the broadcasters are users rather than owners of the UHF bands, it was decided that the duty to manage the white spaces should rest with Ofcom, who would thus need to develop enabling regulation.

Ofcom formed an industry/regulator working group chaired by its former Head of R&D, Professor William Webb. The group comprised representatives of existing users (broadcasters and PMSE manufacturers) and new technology providers (Microsoft, Google, etc.). The group looked at different models for enabling access to the TV white spaces, starting with cognitive radio techniques. It quickly became clear that only the use of geolocation combined with frequency databases could calm the nerves of existing users about the vulnerability of their coverage. This was the same conclusion that the FCC had reached in its proceedings on TV white spaces in the United States.

By 2010, Ofcom had developed a credible and practical approach to enabling licence-exempt access to the TV white spaces. However, it felt that implementing the detailed and novel regulation would take a considerable effort and it was unsure of the industry appetite for the new spectrum access that would result. The Cambridge White Space Trial – 2011–2012 produced a significant amount of technical data and also clearly demonstrated industry's strong interest in TV white spaces.

Ofcom made the decision to progress its work on the regulations. A pilot of the draft regulations was conducted between 2013 and 2014 – leading to a statement [2] of Ofcom's plans for the final regulations. The resulting regulations entered into force at the end of December 2015. According to the new regulations, devices meeting the regulatory requirements (laid down in ETSI EN 301 598 [5]) and used in conjunction with an approved geolocation database were exempt from licensing [3].

However, since the only TV white space devices did not support geolocation – an essential requirement of 301 598 – Ofcom introduced a Manually Configured Device Licence [4].

This licence is available to operators which are qualified to configure TV white space devices in accordance with the licence conditions. The key parameters concerned are the location where the device is installed and details of its antenna – neither of which is integral to the device, but material to ensuring protection of other licensed services in the band.

## 1.7 UK INDUSTRY INTEREST IN TVWS GREW IN PARALLEL WITH REGULATORY DISCUSSIONS

Just as in the United States, the limitations of rural access to the Internet is a significant political issue. Whilst services such as telephony and television were available to nearly 100% of homes, broadband was not as widely available.

A key reason for this was that the predominant means of delivering broadband access has used Digital Subscriber Line (DSL) technology over existing telephone lines. The performance achieved with DSL is highly dependent on the length of telephone line between the exchange (or cabinet) and the end user/customer premises. Whilst the technology could deliver acceptable performance for many urban users, the longer line lengths in rural areas greatly reduced its efficiency.

### 1.7.1 Rural Broadband Deficit Was a Driver of Interest in TVWS

Wireless technology was a clear alternative to DSL technology, enabling rapid deployment of broadband connectivity and better performance in many cases. Much of the available wireless technology had been developed for frequencies of 2.4 GHz and above (e.g. 5 GHz), where there was already spectrum available on a licence-exempt basis in many locations around the world. The only problem with these frequencies was that the range was limited at the allowed emission power, typically 100 metres or less. Going to lower frequencies enables greater coverage for the same power and helps reduce infrastructure deployment costs.

In the early 2000s, major technology providers had started to look at the possibility of licence-exempt access to sub-1 GHz bands. There was little or no clear spectrum available that was not already assigned and licensed. However, geographical and temporal usage was a very different story. Studies by industry and regulators [1] had demonstrated that only a small fraction of available spectrum was in use at any given place and time.

The UHF bands were a clear example of underuse, and where there was the greatest potential for harmonisation around the world, albeit with some regional variations. The harmonisation potential is key to driving new technology costs down and encouraging more widespread adoption. The unused spectrum was referred to as TV white spaces and it was clear that its use could bring major coverage improvements. Key players in the technology sector then set about the business of convincing regulators.

In the UK, regulatory measures had led to a competitive residential broadband market with choice and value in urban areas. However, many in rural areas (as was also true in the US) were feeling left out of the digital revolution. This was ironic, as rural residents and businesses have arguably more to gain from broadband than their urban counterparts: it can offset the greater travel difficulties in rural areas.

### 1.7.2 Spectrum Scarcity Was Another Driver

In the UK, as elsewhere, the launch of the Apple iPhone in 2007 led to radical growth in data being delivered to mobile devices. Steve Jobs and his colleagues were very successful in putting convenient Internet access in consumer pockets and this was a radical stimulus for demand. Mobile network operators saw their voice-to-data ratio invert over the course of a year or two from 90% voice:10% data to 90% data:10% voice. Interest in having more spectrum available followed.

Major UK communication players were highly interested in how licence-exempt access might be used to serve mobile devices as well as a host of other emerging applications (which can be considered under the 'Internet of Things' umbrella). This interest was particularly acute for major Internet service providers, but also relevant to mobile operators, whose networks were becoming fully loaded in some areas.

It was already clear by 2009 that Wi-Fi was an important tool for serving the nearly insatiable data demands of users of smart mobile devices (including tablets and smart phones). Technology that could harness the TV white spaces held the promise of adding capacity and coverage to Wi-Fi networks.

With growing UK industry interest, Microsoft formed a consortium with a number of leading industry partners to undertake a trial of the TV white spaces technology. The activity was centred on the city of Cambridge, which is the location of the Microsoft UK research laboratory and a recognised international hub for technology innovation (see Section 1.10, in this chapter).

## 1.8 PROVING THE VALUE AND WORKABILITY OF TV WHITE SPACES

Industry interest in TV white spaces had grown with the development of a UK regulatory framework which provided the confidence to invest in evaluating the technology. Broadcasters and others were also interested in discovering any potential negative impact of the new technology on their established distribution networks. The regulators wanted to understand the potential value and the technical issues to be addressed in ensuring coexistence.

With these interests in mind, there were two key trials of TV white spaces, one in Scotland and the other in Cambridge. The experience and results gained from these trials helped Ofcom significantly along the road to implementing the regulations.

## 1.9 TV WHITE SPACES TRIAL ON THE ISLE OF BUTE, SCOTLAND (2011 ONWARDS)

The Isle of Bute can be reached by boat after a short sail down the river Clyde, where the University of Strathclyde is based, in Glasgow. Like many other rural communities, the island had suffered from poor broadband for many years. With the aid of innovation funding from the UK Government, the Centre for White Space Communications led a project to bring much improved broadband to the south end of the island using the latest TVWS technology. The project partners included BT and BBC, and its results can be viewed here [7].

## 1.10 THE CAMBRIDGE WHITE SPACES TRIAL (2011 TO 2012)

The Cambridge White Spaces Trial (CWST) was a watershed in UK TV white spaces development, and it proved influential in Europe and other parts of the world. It came at a critical time, when Ofcom had published proposals for a practical mechanism to enable licence-exempt access to TV white spaces.

Ofcom had laid out its technical proposals for enabling TV white spaces regulation in a Statement in early 2011. It believed that the proposals were practical and could meet the concerns and interests of all the key stakeholders and enabling greater value from the UHF spectrum without impairing any of the existing services and applications.

With growing UK industry interest in the TV white spaces from 2008 onwards, it seemed a good time to conduct a pilot TV white spaces deployment. There were two key objectives:

- Establishing that TV white spaces applications could coexist with the licensed services, digital terrestrial broadcasting and programme making and special events (PMSE);
- Exploring how well TVWS communications could help meet the connectivity requirements of a range of applications.

Microsoft Research in Cambridge was an ideal host for the pilot. As a leading centre for wireless technology innovation, Cambridge was also well placed. It already hosted a number of the companies which were interested in participating. In the end, nearly twenty leading companies were involved with the trial in one way or another, including

- Major ISPs (BT, Sky, Virgin Media);
- Broadcasters (Arqiva and BBC);

- Technology providers (Alcatel Lucent, Adaptrum, CSR, CRFS, Microsoft, Neul, Nokia, Samsung);
- Geolocation database providers (Spectrum Bridge and Microsoft);
- Technology solution developers/providers (TTP and Cambridge Consultants).

Cambridge's compact size enabled a range of rural and urban deployment scenarios, including

- Fixed wireless access (broadband);
- Machine-to-machine communications;
- Local content distribution.

Results and recommendations from the trial are available in [8].

A brief summary of the applications is given below.

### 1.10.1 Improving Broadband in a Rural Location

Rural communities around Cambridge had often suffered from poor broadband connectivity. Instead, city-based industry and academic premises have enjoyed excellent broadband access performance. People working and studying in the city noticed a huge difference in the usability of the Internet when they went back home to their rural villages.

Two companies participating in the trial constructed TV white spaces links of 6 km or more from their well-connected offices to rural village locations. Using directional antennas, the links enabled a radical improvement of the broadband performance (from previous sub-megabit levels to over 10 Mbps). The favourable propagation facilitated by the use of TV white spaces was the key factor in the good performance achieved with this low-power link.

### 1.10.2 Machine-to-Machine Communication

Extending broadband coverage is an important objective for UK policymakers, and TV white spaces technology has much to offer in achieving this affordably. However, it is not only people who need Internet access. Increasingly, a range of devices need to be reachable and be able to reach other devices in order to deliver system benefits. The devices may be sensors from which cloud-based applications can harvest information, air quality for example, where each sensor is located. They might also be actuators or output devices, so that the results of monitoring can be used to adapt traffic flow in an urban centre and thereby modulate the pollution levels.

This emerging distributed system approach has been dubbed the *Internet of Things* (IoT) and holds great prospects for improving living standards and the sustainability of communities. It is being applied in urban areas first (in what are called *smart cities*), but is also applicable to rural areas where the difficulties of physical access favour remote management.

The role of TV white spaces here is to find additional spectrum capacity for this new application permitting cost effective coverage across wide areas. Databases provide a natural service interface for devices, so obtaining frequency availability information from a geolocation database seems quite logical. Instead of having frequency use hard-programmed into devices, based on dedicated spectrum regulators can optimise the sharing of spectrum through the database and adapt the sharing rapidly according to changes in market needs according to time and geography. So, a tractor on a farm in the Outer Hebrides need not have the same constraints in linking to the farmhouse and nearby sensors that would be necessary to protect other users on a similar device in the theatre district of central London.

### 1.10.3 Local Content Distribution

TV white spaces technology can deliver a considerable range within power limits that are quite low. Even with the low-power early-generation TVWS products available in 2011, it was possible to achieve distribution bit rates of megabits per second over a range of 1–2 km from an access point, using an omnidirectional antenna at the base station. This made the technology interesting for local content distribution. We looked at the potential to distribute audio-video content from a theatre where one of the TVWS base stations was located.

### 1.10.4 Location-Based Services

Devices within range of a TV white spaces base station/access point can receive information which is relevant for that location.

The granularity of location services can be adjusted by the density of the network deployment, e.g. more access points serving smaller areas. Although TVWS technology can provide much wider area coverage than technologies using higher frequencies, its range can be reduced through reductions of transmitted power and antenna height. Using a geolocation database to provide location-based content might seem easier, but this depends on being able to determine the position of devices, which is more complex indoors. Under the UK regulations, only TVWS base stations or master WSDs need to have direct access to the database and know their precise position. The client or slave WSDs just need to communicate with the master device; they themselves do not need to connect to a database.

In the Cambridge Trial, Duxford Air Museum (which is part of the Imperial War Museum) hosted a demonstration of the power of TV white spaces technology to enhance visitor experience: a variety of relevant multimedia content was delivered depending on which aircraft the user was approaching.

### 1.10.5 Input to Regulatory Working Groups

An extensive technical programme accompanied the trial, allowing evaluation of the potential impact of TVWS communications on existing services: terrestrial TV, cable and UHF wireless microphone links.

The results of these tests were fed into Ofcom's technical working group to assist the development of UK regulations, and also into the Conference of European Postal & Telecommunications (CEPT) working groups (SE43 and later FM53) and ETSI who were developing a harmonised approach for Europe.

## 1.11 OFCOM'S TV WHITE SPACES PILOT (2013–2014)

When Ofcom had completed the first draft of the UK TVWS regulations, it wanted to understand how they would perform in real applications and network deployments. It needed to establish whether the regulations were too cautious in protecting the established services, or not cautious enough (as TV band incumbents had been arguing).

Ofcom invited applications for pilot projects which would benefit from its pilot TVWS licence. Successful applicants delivered a range of projects, including

- Bringing Wi-Fi onto ferries in the Orkneys;
- A video link at London Zoo offering insights into meerkat lives;
- IoT image sensor backhaul and Wi-Fi coverage on Strathclyde University's city-centre campus in Glasgow. There were also tests on indoor TVWS networks in domestic properties of different types of construction [9];
- Multiple links and TVWS implementations around Kings College, London;
- Flood monitoring around Oxford using distributed sensors with TVWS backhaul.

## 1.12 FOUNDATION OF THE CENTRE FOR WHITE SPACE COMMUNICATIONS

The Centre for White Space Communications (CWSC) is now an accepted international authority on developments in TVWS technology and its application [6], particularly to connectivity in remote/rural locations. It has deployment experience in Scotland and in Africa.

It started within the University of Strathclyde, which has a signal processing group within the Electrical and Electronic Engineering department. The group had been working on technology for wireless communications for many years and grappling with the issue of how to bring better broadband to rural areas. Strathclyde had developed a model for powering wireless broadband equipment using solar or wind energy, to enable off-grid Internet connections. It was called 'Wind-Fi'.

Professor Bob Stewart, Head of the EEE Department, and his colleagues were quick to appreciate the advantages of TV white spaces technology in terms of its potential for cost-effective rural coverage. They could see how TVWS expanded the scope of the Wi-Fi technology which had been the mainstay of rural broadband provision up to that point.

Since Wi-Fi was cheap and used licence-exempt spectrum access, it was possible to develop access solutions which would be affordable for relatively poor rural populations. This was particularly important in Africa, but also relevant in rural Scotland and other parts of the UK and Europe.

So, in 2011, Bob and his colleagues established the Centre for White Space Communications within the University of Strathclyde, initially in cooperation with the University of Stirling and the University of the Highlands & Islands. The Centre's foundation was supported by the Scottish Funding Council and by a number of leading industry players.

## 1.13 APPLYING WHITE SPACES: MEETING CHALLENGE OF SCOTTISH RURAL COVERAGE

Scotland has extensive areas of countryside with a relatively low population density. Whilst the Central Belt (Glasgow and Edinburgh and the connecting corridor) houses most of Scotland's population, there are many people scattered across the country's often wild and rugged landscape and on its islands to the west and north. The combination of difficult terrain and low population density makes it hard to build a business case for communications infrastructure to cover everywhere. Broadband and mobile coverage tends to suffer as a result without significant public intervention.

Thanks to major government initiatives, optical fibre is coming ever nearer to smaller and more remote communities; however, it is clear that it would be very expensive and take a long time to serve 100% of properties directly with optical fibre.

In recent years, the Scottish Government contracted British Telecom to extend the availability of faster broadband in rural locations, as has happened in other parts of the UK. However, whilst its roll-out of additional optical fibre has been helpful, most properties still depend on the long-established telephone wire networks for the last stage of their broadband connection. These wires are generally not a good fit with delivering broadband in rural settings, due to the long line lengths and to their age and exposure to weather, etc.

Using wireless links from customer premises to the nearest fibre access point offers a much faster way of delivering affordable and usable broadband access to many more homes and businesses in these communities.

Practical studies conducted by the Centre for Whitespace Communications at the University of Strathclyde, supported by the Scottish Government and undertaken in cooperation with industry, show how TV white spaces technology can assist coverage challenges in rural Scotland, including marine areas. Its coverage efficacy has also been demonstrated in urban areas where applications ranging from social housing connectivity to backhauling sensor data for IoT applications.

Affordability of broadband access is an issue in rural economies where wages in hospitality and food production centres are low in general and also tend to fluctuate with season. To extend coverage to these areas requires a different approach to infrastructure deployment and operation.

## 1.14 A HYBRID ACCESS TECHNOLOGY MODEL IS NEEDED

The diverse nature of the UK landscape, ranging from low flat plains through hilly and even mountainous districts, creates challenges for extending broadband coverage to all. A hybrid wireless networking architecture, for example combining high-capacity 5 GHz with the coverage strengths of TV white spaces, offers a powerful tool for extending broadband access to all, regardless of their surroundings.

Choosing the appropriate technology mix, operators can optimise the performance or value for money they can offer to end users.

- Where line-of-sight wireless links are possible to a fibre hub, 5 GHz Wi-Fi technology offers good performance and value.
- In other cases, hilly or wooded terrain precludes a direct line of sight and favours lower frequency technologies such as TV white spaces.

By making use of a range of frequencies, it is possible to tailor coverage more efficiently to the needs of a given location.

The first commercial roll-out of 'a hybrid 5 GHz / TV white spaces network' – started its roll-out in mid-2016 on the Island of Arran, which offers as challenging a landscape as anywhere else in the UK. The powerful combination of these technologies has proved a good fit with the challenges and has brought radical improvements to broadband access, changing lives and businesses for the better [10].

Urban areas can also benefit from the use of a range of frequencies for broadband access, where high frequencies deliver additional capacity around fibre-fed hotspots in busy venues and lower frequencies (such as TV white spaces bring) help to fill the gaps between the hotspots. Lower frequencies, for example using the IEEE 802.11af (which is part of the Wi-Fi family of standards), can enable more reliable coverage for in-building wireless LAN applications [11].

### 1.14.1 Point-to-Multipoint Architectures Can Lower Deployment Costs

Whereas 5 GHz Wi-Fi technology might need to use a number of sectors radiating from a community hub, a single TV white spaces radio or cluster with an omnidirectional antenna would offer more comprehensive and potentially more cost-effective coverage around a hub because of the use of lower frequencies. This could be helpful where bandwidth demand is lower, for example with a sparser population and/or users who only need occasional access to emails, etc.

## REFERENCES

[1] Spectrum Framework Review Consultation Document, Ofcom, 28th June 2005, https://www.ofcom.org.uk/__data/assets/pdf_file/0014/25403/sfr.pdf.
[2] Ofcom, Statement on implementation of TVWS regulations, Feb. 2015, https://www.ofcom.org.uk/__data/assets/pdf_file/0034/68668/tvws-statement.pdf.
[3] Ofcom, Licence exemption regulations, https://www.ofcom.org.uk/consultations-and-statements/category-3/wireless_telegraphy_white_space_devices_exemption_regulations.
[4] Manually configured device licensing, https://www.ofcom.org.uk/consultations-and-statements/category-2/manually-configurable-wsds.
[5] ETSI Harmonised standard for White Space Devices (EN 301 598), currently v1.1.1, but a new version is expected before the end of 2017, http://www.etsi.org/deliver/etsi_en/301500_301599/301598/01.01.01_60/en_301598v010101p.pdf.
[6] Centre for White Space Communications, University of Strathclyde, http://www.wirelesswhitespace.org.
[7] Bute Trial, http://www.wirelesswhitespace.org/projects/white-space-trial-on-the-isle-of-bute/.
[8] Cambridge White Spaces Trial Recommendations Report, http://download.microsoft.com/download/6/5/1/651469BB-6D84-4B24-A614-8A85E6052543/Cambridge TV White Spaces Trial Recommendations Report.pdf.
[9] In-Home TVWS Measurements Using IEEE 802.11af, D. Crawford, University of Strathclyde, May 2015, http://www.wirelesswhitespace.org/projects/in-home-tvws-measurements-using-ieee-802-11af/.
[10] Arran Broadband, see http://www.arranbroadband.co.uk.
[11] Marine applications of TV white spaces, http://www.baesystems.com/en/article/bae-systems-at-the-heart-of-unmanned-warrior#.

Part I

# Technologies for TV White Space Networks

Chapter 2

# TV White Space Spectrum Sharing Using Geolocation Databases

**Heikki Kokkinen**
*Fairspectrum, H2020 Coherent, Espoo, Finland*

## 2.1 INTRODUCTION

Unused radio spectrum on terrestrial TV band (470–694 MHz) is taken into a new use in TVWS (TV White Space) spectrum sharing. A TVWS geolocation database is an Internet service, which responds to TVWS device queries. The response contains the radio resources, which are available in the device location. Cisco [1] Visual Networking Index (VNI) forecasts 22% compound annual growth rate of Internet traffic globally for years 2015–2020. At the same time, VNI forecasts the Internet traffic from wireless and mobile devices to grow from 48% to 66% of IP (Internet Protocol) traffic in years 2015 to 2020. The growth requires an increase of wireless capacity, which can be realised by improving technology, increasing spectrum, or decreasing cell size in congested areas. TVWS spectrum sharing increases radio spectrum for Internet access by allowing White Space device communication on frequencies, which are currently not fully utilised. Terrestrial TV networks are mainly Multiple Frequency Networks (MFN), meaning that the neighbouring transmitters use different frequencies in order to avoid interference in the TV receivers. In Europe, Digital Video Broadcasting Second generation Terrestrial (DVB-T2) networks support Single Frequency Networks (SFN), where neighbouring transmitters transmit on the same frequency. In most European countries, where DVB-T2 SFN networks are deployed, the main transmitters still operate as MFN relative to other main transmitters. The terrestrial TV frequency plan in Europe follows International Telecommunications Union Radiocommunication sector (ITU-R) GE06 [2] plan. The original allotments define geographically areas and allowed broadcast frequencies in each allotment. The frequencies, which are not used in the allotments are called TV White Space. The amount of TV White Space was increased significantly after the transition from analog

**TV White Space Communications and Networks. http://dx.doi.org/10.1016/B978-0-08-100611-5.00003-5**

to digital TV broadcasting. The final step of the transition was called Analog Switch-Off (ASO).

Both Europe and US support spectrum sharing on the highest political level. In the US, Barack Obama [3] released a presidential memorandum to unleash 500 MHz governmental spectrum to commercial use. A significant part of the spectrum will become shared spectrum. The President's Council of Advisors on Science and Technology [4] responded to the memorandum with a set of recommendations on how to make the required 500 MHz available by facilitating spectrum sharing as a mainline approach. Similarly, the European Commission [5] set a target to develop an European Union (EU) approach to identify beneficial sharing opportunities in harmonised or non-harmonised bands, and to use shared spectrum access rights as regulatory tools to authorise licensed sharing possibilities with guaranteed levels of protection against interference.

TVWS legislation was approved in the US [6] as Second Memorandum Opinion and Order in 2010 and the first TVWS database administrator [7] and devices [8] were approved in 2011 and 2012, respectively. IDA published the Singapore TVWS regulation in 2014 [12]. Ofcom released the regulation for TVWS in the UK in the end of 2015 [9] and approved Fairspectrum as the first TVWS database operator to provide services [10] in the early 2016. On general level, the purpose of the regulations is very similar: to apply geolocation database service to control the radio transmissions of White Space devices so that they do not cause harmful interference to terrestrial TV receivers or Program Making and Special Events (PMSE). The technical implementation differs significantly in the specifications.

Cognitive radio systems generally contain three options for spectrum management: geolocation database, beacons, or spectrum sensing. So far, all TVWS regulations, including the US, the UK, and Singapore, protect incumbent users with a geolocation database. The geolocation database control of radio transmissions is introduced in commercial use for the first time in TVWS systems. Although electronically controlled coexistence of several radio systems has been available using listen-before-talk and on Industrial Science and Medical (ISM) bands, TVWS is often considered as the first commercial realisation of a cognitive radio management system.

Spectrum sharing can be divided in exclusive spectrum use, i.e. no spectrum sharing, static sharing with radio licenses, dynamic sharing, and public access like in WiFi. The coordination of spectrum can be National Regulatory Authority (NRA) coordinated, electronic control like geolocation database or listen before talk equipment, or uncoordinated. The priority levels of the users can be primary, co-primary, secondary or unspecified. The spectrum user may be protected, protected from other but higher priority users, protected from lower priority users, or not protected. The use of spectrum may require a license or it

| Licensing | Interference protection | Priority | Coordination | Sharing |
|---|---|---|---|---|
| Licensed | Protected | Primary | NRA coordinated | Exclusive |
| **License-exempt** | Protected from other but higher priority users | Co-primary | ***Database coordinated*** | Static sharing |
| | Protected from lower priority users | **Secondary** | Listen before talk | ***Dynamic sharing*** |
| | **Not protected** | Unspecified | Uncoordinated | Public access |

**FIGURE 2.1** Classification of spectrum access.

can be license-exempt. TVWS regulation has so far been license-exempt, unprotected, secondary, database coordinated, and dynamically shared. From global perspective, practically all spectrum bands are shared. Regionally or per country, there can be exclusively allocated spectrum bands, but even then more than 50% of spectrum is shared by different type of users. By far, the most common way of spectrum sharing is static sharing. Mostly but not always, radio communication using exclusive radio licenses is protected from harmful interference by the radio administration. License-exempt use is not interference protected, and dynamic spectrum sharing can be used to provide coordination for both interference protected and unprotected radio spectrum. In the US, the UK, and Singapore, TVWS radio communication is unprotected and license-exempt with the exception that the manually configurable devices require an unprotected radio license in the UK and Singapore has High Priority Channels. Between licensing models and sharing types, we can recognise different ways of coordination. Radio licenses are the typical way of spectrum coordination for a radio administration. On certain bands, the radio licenses may be required but the mutual interference coordination is carried out by the industry. For example, when PMSE bands require a license, the coordination can be industry coordinated with the exception of the very large events. A TVWS geolocation database can coordinate the radio use between the primary and secondary users, but it could also be used to coordinate the transmissions of the radio users with the same priority level. Listen before talk equipment can coordinate transmissions locally. One of the most common uncoordinated spectrum use for general public is Industrial Scientific and Medical (ISM) band, which is used, for example, by WiFi and Bluetooth. Fig. 2.1 shows the relationship of different licensing, sharing, and coordination models in spectrum management.

In this chapter, we discuss the TVWS geolocation database from the following aspects. What is the role of a geolocation database as a regulative spectrum

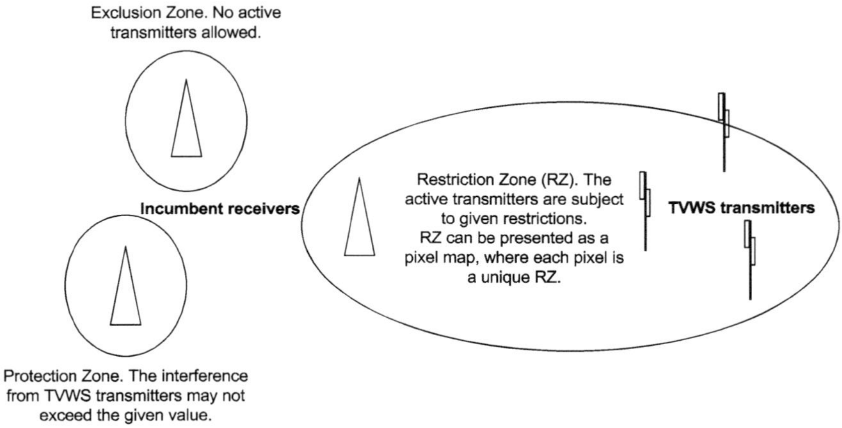


**FIGURE 2.2** Interference protection types.

management option? What is needed from the incumbent radio systems and TVWS systems, which share the spectrum with a geolocation database? What are the characteristics of a spectrum sharing arrangement where a geolocation database can help? Which are the interfaces to interact with a TVWS geolocation database? We give an example of TVWS geolocation database regulation, and the technical implementation of the regulative rules by Fairspectrum, and finally we conclude the chapter by summarising the key points and estimate the role of a geolocation database in the future.

## 2.2 ROLE OF A GEOLOCATION DATABASE

The main purpose of a TVWS geolocation database is to ensure interference free communication for the protected radio users. The radio administration has a legislation based responsibility to regulate the radio transmissions. The radio administration decides what are the best methods in each specific case. If the radio administration plans to apply a TVWS geolocation database, they determine the protection criteria for the TVWS system to avoid harmful interference to protected radio systems. The criteria can generally be described with three types of zones. They are called Exclusion Zone (EZ), Restriction Zone (RZ), and Protection Zone (PZ). Active transmitters are not allowed in Exclusion Zone. In Restriction Zone, the transmitters have a limited operating parameters like maximum transmit power, antenna height, and frequency range compared to the generic transmission conditions. A Protection Zone defines what is the maximum allowed interference level caused by the transmitters inside the Protection Zone. A visual presentation of EZ, RZ and, PZ can be found in Fig. 2.2. Interference protection with Exclusion Zone and Restriction Zone is simplest implemented by checking if the WSD is inside or outside the zone area. The Protection Zone requires propagation model computation, and it is generally more complicated for the geolocation database provider, but at the same time it is also more rewarding than EZ and RZ. The Protection Zones allow more efficient use of spectrum in dynamic conditions, because they can take into account the dynamic changes in aggregated interference from WSD network and on the propagation path. If the conditions differ from location to location, it is very difficult incorporate the rules in the manufactured devices and even in the radio configuration settings without professional understanding of the radio system. The task becomes close to impossible if the protection needs to change rapidly. The geolocation database automates the rules and regulations of the radio administration to fulfil the protection criteria. Automated, electronic control also makes it possible to adapt to fast changes in protection requirements. When there are several geolocation database providers, like in the US, the UK, and Singapore, the functionality of a geolocation database has to be specified and tested very detailed. The tight control of the geolocation database ensure that the responses of all geolocation databases are very deterministic and possible error situations can accurately be tracked. The control also has an effect that future improvements become more complicated because other than parameter modifications require contractual changes with all database providers, detailed specification work, and completed testing procedures. When the radio administration has only one geolocation database, the level of specification could be left more generic and comparable to the level of the normal the radio licenses, leaving the geolocation database provider more room for future improvements and making it more straight forward for the radio administration to implement changes.

In addition to interference protection in the US and Singapore TVWS, the geolocation database collects and maintains incumbent (PMSE) information. Collecting and maintaining incumbent and TVWS device information can be an added role for a TVWS geolocation database in most jurisdictions and radio administrations. Several other functions have been proposed for a TVWS geolocation database: management of radio licenses, transfer of radio license rights, and coexistence management of TVWS device transmissions. In most countries, only radio administration is allowed to issue radio licenses. Due to that, TVWS geolocation database can most likely issue radio license only if it is managed by the regulator. In a few countries, the radio licenses contain rights to transfer or sell the radio license to another organisation. In those countries, it is feasible to consider that a TVWS geolocation database could facilitate sales of radio licenses. If TVWS geolocation database manages radio communication, which requires licenses and licenses offer protection, it is natural that

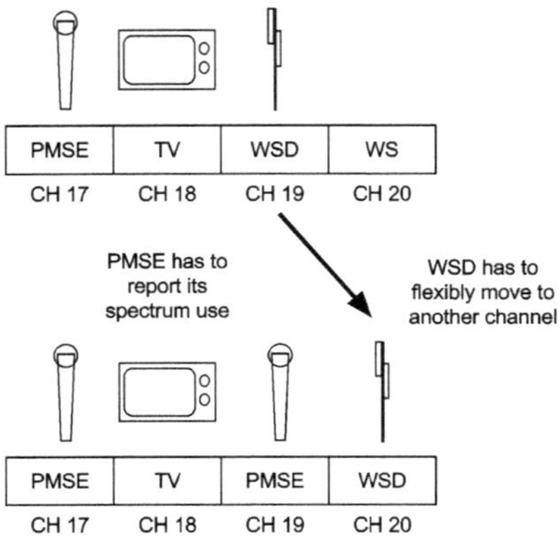


**FIGURE 2.3** WSD flexibility in Dynamic Spectrum Access.

the TVWS geolocation database takes care of the coexistence management of the license holders. In the US, the UK (excluding manually configurable devices), and Singapore (excluding High Priority Channels) TVWS devices are operated under license-exempt regime. In license-exempt regime, the radio administration does not provide protection. If coexistence management is used, it is more likely to be an own responsibility of the users than a service from the regulator. Nevertheless, the regulators could be interested in coexistence management between the radio users on the same priority level provided by the geolocation database. The primary method in coexistence management is basically to limit the transmit options in time, frequency, and power to ensure that the existing users do not experience harmful interference. Authorising some users to transmit and not authorising others is getting close to regulative authority of the radio administration and at least at the moment it would be a big change in legislation and regulation in many countries. If such a geolocation database service is provided by the regulator, then the legislative and regulative challenges are much smaller than if the authority is given to an external organisation.

## 2.3 WHAT IS NEEDED FROM ORGANISATIONS WHO ARE SHARING

The incumbents have to inform the TVWS geolocation database when, where, and on which frequency they communicate. TVWS devices have to be able to tell the geolocation database their location, to communicate with the geolocation database, and to change their operation frequency and power levels according to the geolocation database instructions. Fig. 2.3 shows a simple case how the primary users report the spectrum use and White Space Devices flexibly accommodate the changes in spectrum use. On TVWS frequency range, the most common incumbents are Digital Terrestrial TV (DTT) and Program Making and Special Events (PMSE) like wireless microphones and in-ear monitors. DTT information changes rarely and it is provided by the radio administration in the US, the UK, and Singapore. In the US and Singapore, the DTT data is the actual transmitter information, e.g. latitude, longitude, center frequency, transmit power, and antenna height. Based on the transmitter information, the TVWS geolocation database computes where the TV receivers can be located and most importantly where and on which frequencies the TVWS devices can transmit without causing harmful interference to the TV receivers. In the UK, the radio administration uses the residential house locations to take into account the TV receiver locations. The radio administration computes the maximum TVWS transmit power on each TV channel per 100 m × 100 m pixel. The pixel maps are computed for a number of parameter combinations. The disadvantage for pre-computed data in the UK is that radio administration has to make a significant effort to process the data. It also makes it difficult for a geolocation database provider to help TVWS operators by computing the channels with least interference when the geolocation database provider does not have the DTT transmitter information. The amount of pre-computed data which has to be transferred from the radio administration to TVWS geolocation database provider is huge, making frequent changes in the DTT information impractical. The advantages are that using the receiver locations, the computation is much more accurate than by just using the transmitter information and providing the TVWS device maximum transmit power levels and the incumbent does not have to reveal information about its network to the geolocation database provider. Radio administration can also mix requirements from any other protected systems like public safety or military into pre-computed data without revealing even the existence of such systems.

In the US, PMSE information partially comes from radio administration and partially directly by the user entries in a TVWS geolocation database. The user entries are synchronised between the TVWS geolocation databases. In the UK, PMSE information comes from the radio administration to TVWS geolocation database. In both cases, PMSE information is basically the transmitter location and power. In the UK, there is additionally a PMSE venue polygon describing an area where PMSE equipment can be located.

## 2.4 WHEN A GEOLOCATION DATABASE CAN HELP

The most common alternatives to geolocation database are static licenses and listen before talk on licensed exempt or on licensed industry coordinated band.

Licensed and protected radio communication is typically protected with static licenses. The use of static licenses requires that the protected use does not vary too often, i.e. at maximum few times per year. It also requires that the radio users are knowledgeable about the use of radio transmitters, and that they can be reached with a reasonable effort. TV broadcast networks can generally be protected by static licenses, as the changes in the network are rare. The networks of professional communication operators like mobile operators can be regulated by static licenses. The operators understand well the current laws and regulations of radio communication, they are knowledgeable in technical configurations of the radio communication equipment, and in the case of changes in protection need, they can easily be reached individually. On license-exempt band, the radio communication is not protected. The user with the highest transmit power and shortest distance communicates reliably, but the other users must look for an alternative channel for radio communication. Listen-before-talk functionality in the devices can help to avoid interference. A licensed band, which is in common use by a limited number of users, is protected from interference from other types of users by the regulator. The licensed users coordinate the use among themselves. Wireless cameras for example can have common use on the 2.3 GHz band. Listen-before-talk also avoids interference on licensed common use bands.

There are three main situations, where a geolocation database becomes useful: 1) primary and secondary users share the band and the radio use of the primary changes frequently, 2) secondary users are consumers or corporate users, which do not have professional communications staff, and 3) sharing between co-primary users, which either change their use frequently or which are high in number. TV broadcasting and PMSE have a higher priority than TVWS users. The time and location of PMSE use changes frequently. In 2017 most TVWS users are communication professionals, but there is a possibility that the consumers are able to purchase TVWS access points. In that case, it would not any more be possible for the regulator to guarantee that each TVWS access point owner reads the TVWS regulations and configures the access point radio frequencies correctly. When the common use bands become congested and the number of organisations using the band increases, a geolocation database can be a solution for spectrum coordination of that band, like in the Netherlands on 2.3 GHz. A part of the problems can be solved by listen-before-talk, but the geolocation database can provide more control and priority levels within the common use license holders and individual radio transmitters than listen-before-talk.

## 2.5 INTERFACES OF A GEOLOCATION DATABASE

The TVWS geolocation database has machine and human user interfaces. The machines connected to the database include White Space Devices (WSD), incumbent devices, and regulator applications. The geolocation database must be able to retrieve map, topography, clutter, and other data from external servers, and there might be a requirement to share and synchronise data with other geolocation databases.

The White Space Devices (WSD) communicate with the geolocation database using IETF Protocol to Access White Space database (PAWS) protocol [11]. The PAWS protocol communicates in a client-server mode, where the geolocation database operates as a http server and the White Space Device as a http client. The messages are encoded in JavaScript Object Notation (JSON). The JSON messages are transferred using http over TLS (HTTPS). HTTPS has server authentication and client authentication is optional. In the protocol, the WSD access points are called masters and Customer Premises Equipment (CPE) are called slaves. The masters always communicate directly with the geolocation database. The slaves may communicate with the geolocation database, but most often the master, to which slave is connected to, interfaces with the geolocation database on behalf of the slave. The procedures defined in the protocol include: Initialisation, Device Registration, Available Spectrum Query including Spectrum Use Notification, and Device Validation. In the initialisation phase the master device exchanges capability information with the geolocation database. With the registration procedure, the WSD conveys the required registration information to the geolocation database. Available Spectrum Request (ASR) tells the geolocation database that the WSD would like to know what spectrum it could use for communication. A variant of ASR is Available Spectrum Batch Request, which can be used to query the spectrum at multiple locations. With Spectrum Use Notify, the WSD informs the geolocation database about the spectrum anticipated to be used. A master device uses Device Validation Request to determine which slave devices are permitted to operate. In the regulations, where there are several geolocation database providers and the database providers may collect incumbent information, for example about PMSE, or if the regulation specifies coexistence coordination for the TVWS devices, the geolocation databases should synchronise such information between themselves over a machine-to-machine interface.

The incumbent information is inserted either manually through a human user interface or it is collected by the regulator and downloaded to the geolocation database. When the geolocation database use becomes more common, it can be expected that the incumbent devices will be able to report the information about

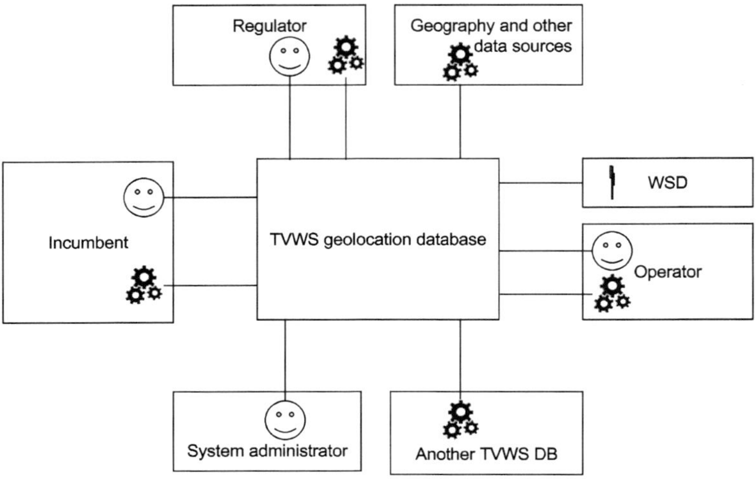


**FIGURE 2.4** Interfaces of geolocation database.

their use automatically to the geolocation database. Ofcom has developed an application for managing TVWS geolocation databases. The application uses White Space Information Platform (WSIP) protocol to send data requests simultaneously to all databases [9]. The WSIP responses include, for example, the location of the devices at the requested time, the used channel and transmit power level. Through WSIP Ofcom can also apply restriction to the devices like frequency range and maximum power level, or cease the device transmission.

The human user interfaces may be needed for device operators, regulators, incumbents, and system administrators. The device operators insert their own information and contact information into the geolocation database. The database allows the WSDs to register through PAWS protocol, or they configure the device registration information through the user interface. In Ofcom TVWS regulation, the license-exempt devices are expected to be able report their location automatically to the TVWS database. When that is not possible for example due to missing GPS module or indoor installation, the devices can be used under manually configurable devices licenses. The information required from the manually configured devices is called device installation record. The information in the installation record is provided through the operator interface before the manually configurable WSD is taken into use. A part of the installation record information can also be configured in the device and provided through the PAWS interface. An alternative to a machine-to-machine (M2M) interface, like Ofcom WSIP, is to provide a human user interface for the regulator to inspect and control the operation of the geolocation database. The generic functions of the regulator interface can include: managing end users, technical parameters, usage restrictions, and log information. The system administrator user interface, which can also be implemented as a M2M interface e.g. with Simple Network Management Protocol (SNMP), is used to manage user rights and authorisation on system level. The administrator takes care of processing environment capacity and status, reacts to alarms, ensures that incumbent information retrieval, billing, system logging, and backup processes work as planned. The administrator may deploy updates both on processing environment and on geolocation database application. Fig. 2.4 shows an example of TVWS geolocation database interfaces.

## 2.6 CURRENT SOLUTIONS

As an example, we study Fairspectrum implementation of the UK TVWS geolocation database. TVWS extends from 470–698 MHz in the UK. The incumbents are Digital TV Transmission (DTT) and PMSE. DTT information is maintained by Ofcom. The protection method is 100 m by 100 m pixel map based Restriction Zones, which define the maximum allowed transmission powers for various WSD parameter sets. Ofcom takes into account known TV receiver locations and residential homes, when computing the DTT protection. Another noteworthy issue, which is taken into account in the DTT protection, is GE06 coordination threshold. It defines Protection Zones, including an agreed antenna height and maximum caused interference, at the border of any country in the GE06 plan. If the regulator expects that the interference exceeds the GE06 coordination threshold value of another country, the regulator must begin coordination negotiations between the countries. As the coordination between the countries takes time and is a heavy process, Ofcom takes the GE06 threshold into account in DTT pixel maps. The consequence is that the maximum power levels in the DTT maps decrease when getting close to a seashore from inland or towards Irish border in Northern Ireland. The task of the database is to select which pixel map to use, which pixels to take into account, and to select the lowest transmit power value for each TV channel. The pixel map selection is done based on the WSD parameters, like antenna height and the RF quality of the transmitter. The pixel selection is based on the WSD location, location inaccuracy, and specific vs generic request type. In specific requests, the slave device has an accurate location, whereas in generic requests the exact location of the slave device is not known. In a generic request, all DTT pixels where a slave is expected to be able to receive the transmission of a master are taken into account. PMSE protection is carried out as Protection Zone represented by corner points of 10 m by 10 m pixels around each PMSE device or PMSE venue. The PMSE information is collected and provided by

Ofcom, but in contrast to DTT protection, the protection computation including the radio propagation models is done in the geolocation database. DTT data changes every 6–24 months and PMSE data every 15 minutes. In addition to DTT and PMSE protection, Ofcom is able to change the maximum TVWS transmit power level at any pixel using unscheduled adjustments. This can be used for example to shut down all TVWS device use around big international events. The license regime in the UK is based on ETSI [13] harmonised standard for WSDs and the UK specific laws and regulations. WSDs are license-exempt devices, when they are ETSI compatible. A specific waiver for ETSI requirement that WSD must be able to geolocate, is manually configurable devices specification. The manually configurable devices require an organisation specific license.

As an example about a geolocation database implementation, Fairspectrum geolocation database is deployed in Amazon EC2 cloud application platform. The service is accessible through load balancers. The Ubuntu linux application servers accept traffic only from load balancers. The databases run in Relational Database Service (RDS) instance, which only accepts traffic from the application servers. The system can send emails through a Simple Email Service (SES). All main computing resources are monitored with CloudWatch. Auto Scaling can be used to adapt to service demand. Inside an application server, a Java client periodically or on request downloads the incumbent and other computation related data to the application server and inserts the data into RDS database. The interference computation is implemented in C++ language, and it utilises PostGIS database extension for geographical queries and Geospatial Data Abstraction Library (GDAL) for geographical function calls. The user interfaces for administrators and operators, PAWS and WSIP protocol interfaces are implemented with PHP. PHP also maintains the main operative logging function. The processing environment architecture of Fairspectrum geolocation database server system can be found in Fig. 2.5.

## 2.7 CONCLUSIONS

Although geolocation databases have successfully been applied for radio spectrum management in several countries, they still are a young option for radio regulation. As a regulative tool, they are between radio licences and license exempt use on Industrial Scientific and Medical (ISM) band. Geolocation databases can manage licensed, industry coordinated bands, and license exempt bands. The standardised systems for radio spectrum geolocation databases are TVWS, Licensed Shared Access (LSA), and Citizen's Broadband Radio Service (CBRS). In addition to these standard systems, a customised geolocation can be used on

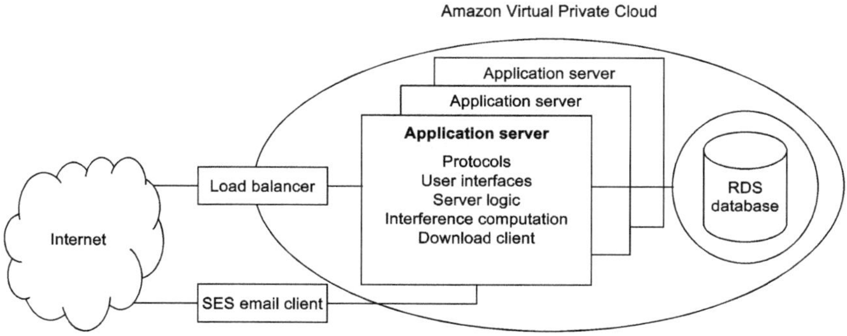


**FIGURE 2.5** Geolocation database processing environment.

most bands, which either need dynamical spectrum management between primary and secondary users, or coexistence management between users with an equal priority level.

The basic requirement for the geolocation based spectrum management is that the protected systems must report the time, location, and frequency of their spectrum use. In optimal case, the reporting is automatic from the devices or from the device management system. The alternative is that the system operator manually informs the geolocation database system directly or through the regulator. The new users, WSDs in TVWS, must be able to adapt their radio use to the geolocation database control. The devices or device management system must be constantly connected to the geolocation database and change the transmission frequency or power according to the geolocation database instructions.

Geolocation database spectrum management can help when there are users with different priorities, and the higher priority use changes frequently. Manual control for following constantly the changes in incumbent use and configuring the required changes to the secondary radio system is impractical. Listen-before-talk works for most systems, which do not have different priority levels, but priority levels are more convenient with a geolocation database. When non-trained consumers are allowed to operate the access points, which determine the frequency to be used, and there are protected users, automated control with a geolocation database is required. On common use bands both listen-before-talk and geolocation database control work, but with geolocation database it is possible to make reservations, have priorities, and schedule the use of frequencies more conveniently.

Ofcom TVWS regulation protects primarily terrestrial TV receivers and PMSE use. The GE06 coordination threshold protection for neighbouring countries is taken into account in DTT data. Any other protection need can also be added by utilising unscheduled adjustment data. The license-exempt WSDs are expected to be ETSI EN 301 598 [13] compliant. For the devices, which cannot geolocate, there is possibility for Manually Configurable WSD licenses. Fairspectrum is one of the Ofcom approved database providers. Fairspectrum geolocation database is implemented in Amazon EC2 cloud service using perimetry security with load balancer, application server, and database security zones.

The development cycle of new radio systems fastens continuously. It is difficult to clear any existing spectrum band for the new systems. Static sharing has already been applied widely. In order to accommodate even more systems on the same band, the pressure to apply dynamical spectrum sharing with geolocation databases inevitably increases.

## REFERENCES

[1] Cisco, The Zettabyte Era: Trends and Analysis, July 2016, White paper, http://www.cisco.com/c/en/us/solutions/collateral/service-provider/visual-networking-index-vni/complete-white-paper-c11-481360.html.

[2] ITU-R GE06, Plans for VHF/UHF analogue and digital broadcasting in parts of regions 1 and 3, in the frequency bands 174–230 MHz and 470–862 MHz, Geneva 2006 (GE06), http://www.itu.int/en/ITU-R/terrestrial/fmd/Pages/ge06-list.aspx.

[3] Barack Obama, Presidential Memorandum on Unleashing the Wireless Broadband Revolution, Press release, The White House, Office of the Press Secretary, June 28, 2010, https://www.whitehouse.gov/the-press-office/presidential-memorandum-unleashing-wireless-broadband-revolution.

[4] PCAST 2012, Report to the president. Realising the full potential of government-held spectrum to spur economic growth, Executive Office of the President, President's Council of Advisors on Science and Technology, July 2012, available at http://go.usa.gov/k27R (PCAST Report).

[5] European Commission, Communication from the Commission to the European Parliament, the Council, the European Economic and Social Committee and the Committee of the Regions, Promoting the shared use of radio spectrum resources in the internal market, Brussels, 3.9.2012 COM(2012), 2012, 478 final, http://eur-lex.europa.eu/LexUriServ/LexUriServ.do?uri=COM:2012:0478:FIN:EN:PDF.

[6] FCC, Second Report and Order and Memorandum Opinion and Order, ET Docket Nos. 02-380 and 04-186, 2012, 23 FCC Rcd 16807 (2008).

[7] FCC 2011/12, DA 11-2044, December 22, 2011, Approval is hereby granted for Spectrum Bridge, Inc. to operate its "TV bands database system" to provide service to the public on or after January 26, 2012, https://apps.fcc.gov/edocs_public/attachmatch/DA-12-113A1_Rcd.pdf.

[8] FCC 2012/4, DA 12-620, Released: April 19, 2012, Office of Engineering and Technology Permits Telcordia Technologies, Inc. to Provide TV Bands Database Service in Nottoway County, Virginia.

[9] Ofcom 2015/2, Implementing TV White Spaces. Statement, 12 February 2015, https://www.ofcom.org.uk/_data/assets/pdf_file/0034/68668/tvws-statement.pdf.

[10] Ofcom 2016/2, White Space Database Operators, List of WSDBs qualified to operate in the United Kingdom and currently providing WSDB Services.Updated on Feb 1, 2016. Available at https://tvws-databases.ofcom.org.uk.

[11] IETF 2014, Internet Engineering Task Force (IETF), Request for Comments: 7545, Protocol to Access White-Space (PAWS) Databases.

[12] IDA 2014/6, Decision paper issued by the Info-Communications Development authority of Singapore regulatory framework for TV White Space operations in the VHF/UHF bands 16 June 2014.
[13] http://www.etsi.org/deliver/etsi_en/301500_301599/301598/01.01.01_60/en_301598v010101p.pdf.

Part II

# Policy, Regulation and Standardisation Issues

Chapter 3

# Indoor–Outdoor TV White and Gray Space Availability: A U.S. Case Study

**Xuhang Ying, Farzad Hessar and Sumit Roy**
*University of Washington, Seattle, WA, United States*

## 3.1 INTRODUCTION

A global transition from analogue to digital over-the-air (OTA) TV broadcasting is underway [1]. Finland was an early leader in terminating all analogue-to-air TV transmission (in 2007), while in the United States, a similar mandate from the Federal Communications Commission (FCC) was accomplished by June 2009. The incentives for this are many: improved picture quality (due to high definition programming, higher resolution digital TV receiver displays, a significant reduction in 'ghosting' etc.), enhancements such as encryption (for pay TV) and electronic information (guides, on-screen display), user interactivity (more on this later), multiple digital streams per analogue channel (due to benefits of digital compression that requires less bandwidth for the same quality), and potential for better coverage (due to single frequency network architecture using lower power DTV transmitters) [2].

The transition to digital terrestrial television freed up 108 MHz (television channels 52 through 69) of spectrum for other uses. These channels will be compressed into the space currently occupied by channels 2 through 51. Some of the freed space is reserved by US legislation for a nationwide public safety network for use by First Responders (police departments, fire departments and emergency/medical personnel). The remaining frequencies are being auctioned off to improve wireless data services in next-generation networks in the 700 MHz band. The spectrum at a given location within channels 2–51 which is not locally in use by any licensed or *primary users* (PUs) deemed worthy of protection (as defined by the US regulatory authority, FCC), is called the *TV White Space* (TVWS) [3]. These PUs include low power transmitters such as boosters/translators/Class A stations, land mobile radio services, wireless microphones and any other incumbents. As can be expected, the FCC has

TV White Space Communications and Networks. http://dx.doi.org/10.1016/B978-0-08-100611-5.00005-9

sought to achieve the reasonable median in their rulemaking process regarding TVWS by balancing the incentives/opportunities for proponents seeking to develop viable new unlicensed or secondary networks based on *TV band devices* (TVBDs), which are considered unlicensed, or *secondary users* (SUs) with measures to ensure that these networks must protect PUs that operate in the TV bands.

### 3.1.1 A Historical Review on TV White Space Ruling in the United States

On May 13, 2004, the FCC issued the Notice of Proposed Rule Making (NPRM), 71 FR 66897, November 17, 2006, which first proposed to allow secondary operations in TVWS. To ensure that no harmful interference would occur to TV receivers and other authorised users of the spectrum, the FCC proposed that the conditions under which a TV channel can be used by SUs be defined, i.e. that secondary devices should be required to incorporate 'smart radio' features whereby they first identify the unused TV channels in the vicinity for permissible use. The unlicensed TVBDs were classified into two functional categories: (a) lower power personal/portable unlicensed devices such as Wi-Fi cards in laptop computers, and (b) higher power fixed unlicensed devices that could be used to provide commercial services such as 802.22 base station/CPEs (Customer Premise Equipment) providing a backhaul for broadband Internet access. The FCC specifically sought comments on the use of spectrum sensing to identify vacant TV channels.

In November 2008 [3], the FCC formally approved secondary operations in TVWS subject to the following broad requirement: *all unlicensed fixed TV band devices and all personal/portable devices, except for those that operate in Mode I under control of a fixed device or in Mode II under control of a personal/portable device, will be required to access a TV band database to obtain information on the available channels at their location, and all unlicensed fixed TVBDs will be required to register their operations.*

In December 2010, the FCC finalised the rules regarding 'Unlicensed Operation in TV Broadcast Bands' [4]. We summarise a list of important requirements for TVBD operations below as a reference for the rest of this work. In this study, we focus on co-channel interference and omit a subset of FCC rules about adjacent channel interference, radio astronomy sites, Personal Land/Commercial Mobile Radio Services and microphone reserved channels.

- Fixed devices: Fixed devices can initiate a network by sending enabling signals to one or more fixed and/or personal/portable TVBDs.
- Mode I personal/portable devices: A Mode I device is not required to use geolocation or have access to the TV bands database. Instead, it obtains a list of available channels from either a fixed or Mode II personal/portable

device. A Mode I device may not initiate a network nor may it provide a list of available channels to another device.

- Mode II personal/portable devices: A Mode II device uses internal geolocation capability and access to the TV bands database either through a direct connection to the Internet or through an indirect connection by way of a fixed TVBD or another Mode II TVBD. A Mode II device may select a channel itself and initiate and operate as part of a network of TVBDs, transmitting to and receiving from one or more fixed or personal/portable TVDBs. It may provide a list of available channels to a Mode I device.
- Permissible channels: Fixed devices may operate on any channel in $\{2:51\} \setminus \{3,4,37\}$, while personal/portable devices may only transmit on available channels in $\{21:51\} \setminus \{37\}$. Channel 37 (608–614 MHz) is allocated to radio astronomy and the wireless medical telemetry service, and is not used for TV broadcasting.
- Power limit: For fixed TVBDs, the maximum power delivered to antenna may not exceed 1 watt in 6 MHz with a maximum of 6 dBi antenna gain (maximum 36 dBm of Effective Isotropic Radiated Power (EIRP)). For personal/portable TVBDs, the maximum EIRP shall not exceed 20 dBm per 6 MHz. If a portable TVBD is transmitting in a channel adjacent to a primary transmitter, then maximum EIRP is limited to 16 dBm.
- Antenna height: The transmit antenna for fixed devices may not be more than 30 metres above the ground. In addition, fixed devices may not be located at sites where the antenna height above average terrain is more than 250 metres. Portable device antenna is assumed to be less than 3 metres above the ground.
- DTV service model: The FCC adopts a model that assumes that TV broadcast signals are received off-the-air by an antenna with 10 dBd gain that is mounted outdoors at 10 metres above the ground (see ¶76 in [3]).
- Interference avoidance methods: Channel availability for a TVBD is determined based upon *geolocation* and *database access*, but *spectrum sensing* can also be used. For fixed and Mode II devices, geographic coordinates need to be determined to an accuracy of $+/-$ 50 metres by utilising incorporated geolocation capability. Fixed devices that are professionally installed do not need geolocation capability if the coordinates are provided at time of installation. All devices must register their coordinates with the database and must re-establish their position each time it is activated from a power-off condition. A fixed device shall access and update its database at least once every day and a personal/portable device must check its location at least once every 60 seconds. Operation on a channel must cease immediately if the database indicates the channel is no longer available. A Mode II personal/portable device must also re-check its position if it changes location during operation by

more than 100 metres. A Mode II device can also establish a bounded area in which a channel is available at all locations within the area and operate on a mobile basis within that area.

- TV bands database information: A TV bands database maintains records for all authorised primary services that operate in the TV bands, and contains the following information: transmitter coordinates, effective radiated power (ERP), antenna height above terrain (HAAT), antenna pattern, beam tilt, channel number and call sign. A TV bands database administrator may charge a fee for the provision of lists of available channels and other associated information and for registering fixed TVBDs.
- Database access: Fixed devices are intended to access the database from the Internet, but those that do not have a direct Internet connection can access the database from another fixed device within the network. A Mode I personal/portable TVBD may only transmit upon receiving a list of available channels from a fixed or Mode II TVBD that has contacted a database and verified that the FCC identifier (FCC ID) of the Mode I device is valid.
- Spectrum sensing: Spectrum sensing [5] may be used in lieu of geolocation and database access, but devices must submit an application for certification and demonstrate with a high degree of confidence that they will not cause harmful interference to incumbents [6]. TVBDs that meet the spectrum sensing requirements for personal/portable devices are limited to a maximum EIRP of 50 mW. All fixed and personal/portable TVBDs must be capable of detecting ATSC digital TV, NTSC analogue TV and wireless microphone signals at the following detection thresholds:
  - ATSC digital TV signals: −114 dBm, averaged over a 6 MHz bandwidth
  - NTSC analogue TV signals: −114 dBm, averaged over a 100 kHz bandwidth
  - Low power auxiliary and wireless microphone signals: −107 dBm, averaged over a 200 kHz bandwidth.

  A spectrum sensing TVBD may start operating on a channel if no primary transmissions are detected within a minimum time of 30 seconds. A TVBD must perform in-service monitoring of an operating channel at least once every 60 seconds to detect a primary; the TVBD must thereafter cease operations within two seconds.

### 3.1.2 TV White Space and Grey Space

As mentioned earlier, FCC rules allow SUs or TVBDs to operate in TVWS without harmful interference to existing PUs, primarily TV receivers (Rx's). This is translated to a *service region* around every TV tower, where TV Rx's are assumed to be everywhere (i.e. the worst-case scenario). As a result, SUs must not

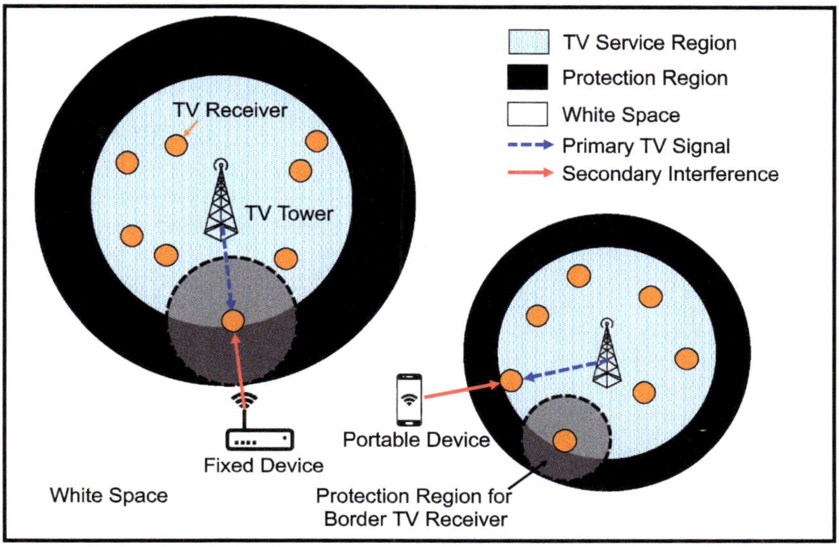


**FIGURE 3.1** Outdoor white space. Both TV towers are using the same channel. For each tower, the inner and outer circles represent its service region and protection region, respectively. If a secondary device is outside the protection regions of all TV towers using the same channel, it can access that channel (which is idle) as a white space. We assume the maximum EIRP for both fixed and portable devices. Note that the protection region for fixed devices is larger than that for portable devices because fixed devices have greater power.

transmit on the same channel within the service region and its vicinity, known as the *no-talk zone*, which jointly form the protection region (see Fig. 3.1) to avoid interference to boundary TV Rx's. Based on the current geolocation and RF parameters, the TV bands spectrum database (e.g. UW SpecObs) determines whether a channel is a WS channel and further provides a list of WS channels for use. UW SpecObs (http://specobs.ee.washington.edu/) is a cloud-based engine for TVWS simulation, maintained by University of Washington (UW) and available for use as a web service.

In contrast, a channel is considered a *grey space* (GS) if the SU is located within the service region, but secondary transmission contributes weak interference to nearby TV Rx's; hance deemed acceptable by the same interference protection criteria defined by the FCC (see Fig. 3.2). As a result, it is necessary to have certain knowledge of TV Rx locations in order to exploit GS opportunities, as assumed in [7]. In this work, we invoke a weaker assumption in determining GS availability, namely that an SU knows the maximum distance within which no TV Rx is present. After GS opportunities are identified, underlay techniques such as spread spectrum (utilized heavily in cellular networks [8]) are potentially available for use.

There is a widespread perception [9–11] that WS channels (and capacity) is plentiful where there is little demand (e.g. in less populated, largely rural areas) and that there is no WS where it is most needed (e.g. dense urban areas). In this study, we propose that the latter is a false 'rush to judgement' from two different perspectives. First, we show that there are more WS channels indoors than outdoors, due to the attenuation effects of walls on secondary interference. Second, we show that there are sufficient GS opportunities outdoors, and many more indoors.

This chapter is organised as follows. In Section 3.2, we define outdoor WS and GS in terms of interference protection. Then we extend the criteria to indoor WS and GS, and provide their definitions in Section 3.3. In section 3.4, we present a case study for San Francisco to compare outdoor and indoor availability for WS and GS, and further understand the contributing factors. Finally, we provide our conclusion in Section 3.5.

## 3.2 INTERFERENCE PROTECTION IN TV BANDS

In this section, we define outdoor WS and GS in terms of interference protection as required by FCC rules.

### 3.2.1 Outdoor White Space

Fig. 3.1 illustrates the concept of outdoor WS. Each TV tower has a service region, where TV signal strength in dBu is no less than a minimum threshold (e.g. 41 dBu for digital services) and TV service is predicted to be available. The electric field intensity unit dBu or dBμV/m is commonly used for signal or field strength in FCC documentation. It is independent of frequency, receive antenna gain and receive antenna impedance, etc. For consistency, we use the unit dBu when referring to signal strength in this study. The relationship between transmit power (mW) and its equivalent signal strength (dBu) is as follows: $E[\text{dBu}] = 10\log_{10} P[\text{mW}] - 20\log_{10} d[\text{m}] + 104.8$, where $d$ is the reference distance (assuming free-space propagation). The equivalent signal power threshold is $-84$ dBm (over a 6 MHz channel). The *protection region* is an area in which SUs are prohibited from accessing the channel, which consists of the service region and the no-talk zone. Formally, WS is defined as *the spatial reuse of a TV channel at locations that are outside protection regions (containing service regions) of all TV towers which are currently broadcasting on that channel.*

We now discuss how to determine the protection region or no-talk zone for interference protection. Denote the received TV signal strength at a TV Rx as $E_{Rx}^{TV}$, which is a function of location and TV transmitter parameters, such as transmit power and antenna pattern. Denote the minimum *desired-to-undesired signal ratio* (D/U ratio) in dB at a TV Rx as $\gamma_{\min}$. For any TV Rx within the service region, it is required that

$$E_{Rx}^{TV} - I_{Rx}^{SU} = E_{Rx}^{TV} - \left[E^{SU} - L_{SU}(d)\right] \geq \gamma_{\min}, \tag{3.1}$$

where $I_{Rx}^{SU}$ is the secondary interference received by the TV Rx, $E^{SU}$ is the transmit signal strength of the SU Tx and $L_{SU}(d)$ is the overall path loss (in dB) between the SU Tx and the TV Rx with a separation distance of $d$.

To use a WS channel, the SU must be outside the service region. Since received TV signal strength decays as a TV Rx gets farther away from the TV tower, a border TV Rx receives the weakest TV signal of strength $(E_{Rx}^{TV})_{\min}$ (e.g. 41 dBu). To protect this TV Rx, the following condition must be fulfilled,

$$E^{SU} - L_{SU}(d) \leq (E_{Rx}^{TV})_{\min} - \gamma_{\min}, \tag{3.2}$$

which specifies the minimum separation distance $d_{\min}$ or protection region for that particular border TV Rx for a given SU Tx,

$$d \geq d_{\min} = L_{SU}^{-1}\left(E^{SU} - (E_{Rx}^{TV})_{\min} + \gamma_{\min}\right). \tag{3.3}$$

As we can see, $d_{\min}$ only depends on SU parameters, path loss and FCC rules, and is independent of PU parameters (including both TV tower and receivers). The union of protection regions for all border TV Rx's (presumably everywhere) excluding the service region defines the no-talk zone for a given SU.

It is clear that a greater $E^{SU}$ leads to a greater $d_{\min}$ and consequently a larger no-talk zone, as illustrated in Fig. 3.1, with the general assumption that path loss monotonically increases with distance. On the other hand, if $d$ is given, Eq. (3.2) provides an upper limit on the transmit power of SU:

$$E^{SU} \leq (E^{SU})_{\max} = (E_{Rx}^{TV})_{\min} - \gamma_{\min} + L_{SU}(d). \tag{3.4}$$

From the perspective of an SU, we can use any of the above equations to check whether it causes interference to the nearest border TV Rx and determine WS availability based on its location and operating parameters such as transmit power.

### 3.2.2 Outdoor Grey Space

In WS sharing, all locations within the service region are protected regardless of actual presence of nearby TV Rx, which leads to conservative and inefficient frequency reuse. Hence, if TV Rx locations are known a priori, we can protect them by prohibiting nearby SUs from transmitting co-channel within their protection regions, as illustrated in Fig. 3.2. As a result, it leads to the notion of

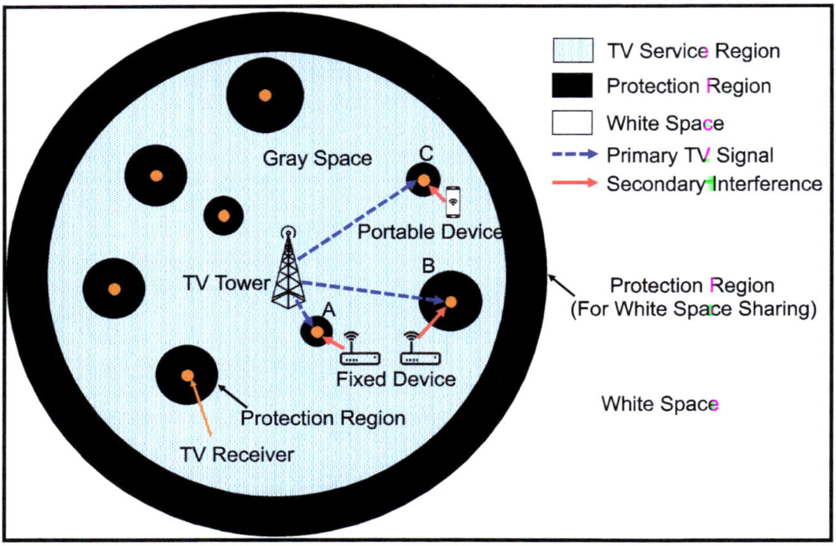


**FIGURE 3.2** Outdoor grey space. The knowledge of TV Rx locations facilitates the determination of protection regions within the service region. A SU that is inside the service region but outside all protection regions can access the occupied channel as a grey space channel. Here, we assume the maximum EIRP for both fixed and portable devices. Note that for TV Rx B and C with the same distance from the TV tower, the protection region for fixed devices is larger than that for portable devices because the former have greater transmit power. In addition, TV Rx A, which is closer than TV Rx B, has a smaller protection region for fixed devices since it receives a stronger TV signal and thus can tolerate greater secondary interference.

GS, formally defined as *the spatial reuse of a TV channel at locations within the TV service region but outside protection regions of all TV Rx's*. Although unlicensed GS operations are currently not allowed, we would like to quantify its availability and explore its potential in improving spectrum utilisation in this study. Note that when a channel is available as a GS inside the service region, it does not necessarily mean it is actually usable for secondary operations. For instance, when the secondary network is very close to the TV tower, it is difficult to reuse the GS channel using underlay sharing techniques (e.g. spread spectrum) because the TV signal, treated as noise, is too strong.

To extend the WS interference protection criteria to GS sharing, we rewrite Eq. (3.1) to account for TV Rx location as

$$\left[E^{TV} - L_{TV}(d_1)\right] - \left[E^{SU} - L_{SU}(d_2)\right] \geq \gamma_{\min}, \tag{3.5}$$

where $E^{TV}$ is the transmit signal strength of the TV tower, and $L_{TV}(d_1)$ is the overall path loss between the TV tower and the TV Rx, which are separated by $d_1$.

For a given TV Rx inside the service region, Eq. (3.5) specifies the minimum separation distance of nearby SUs:

$$d_2 \geq (d_2)_{\min} = L_{SU}^{-1}\left(E^{SU} - E^{TV} + L_{TV}(d_1) + \gamma_{\min}\right). \tag{3.6}$$

If $d_2$ is fixed, Eq. (3.5) then provides an upper limit on $E^{SU}$, namely,

$$E^{SU} \leq (E^{SU})_{\max} = E^{TV} - L_{TV}(d_1) - \gamma_{\min} + L_{SU}(d_2). \tag{3.7}$$

We can see that unlike the WS case, both $(d_2)_{\min}$ and $(E^{SU})_{\max}$ in GS depend on both PU and SU parameters. For PUs, increasing $E^{TV}$ for TV tower and decreasing $d_1$ for TV Rx both lead to less $(d_2)_{\min}$ and greater $(E^{SU})_{\max}$. For SUs, decreasing $E_{SU}$ reduces $(d_2)_{\min}$, and increasing $d_2$ increases $(E^{SU})_{\max}$ effectively.

## 3.3 FROM OUTDOOR TO INDOOR WS AND GS, AND THE IMPACT OF BUILT ENVIRONMENT

When a secondary network is set up inside a building, the building introduces significant penetration loss to wireless signals coming in or going out. According to the COST 231 model [12], the penetration loss can be divided into four major categories: external wall loss, room loss, floor loss and building loss. Due to wall materials and building structures, there can be considerable deviations in penetration loss from building to building. Typical parameters, as recommended by the COST 231 model, are the following: 4–10 dB for the external wall and each internal wall (concrete walls 7 dB), and up to 20 dB additional external wall loss under certain conditions. More details are available in [12].

In terms of interference protection, building penetration loss effectively attenuates secondary interference from indoor SUs, which facilitates frequency reuse. In this section, we discuss the impact of built environments on WS and GS availability.

### 3.3.1 Indoor White Space

According to FCC rules (¶77 [3]), it is necessary for an SU to be outside TV service regions in white space sharing, regardless of its location (indoors or outdoors). Denote the overall *wall penetration loss* due to buildings as $L_{wall}$ (in dB). Similar to the outdoor case, a border TV Rx needs to be protected from an indoor SU,

$$E^{SU} - L(d) - \boldsymbol{L_{wall}} \leq (E_{Rx}^{TV})_{\min} - \gamma_{\min}, \tag{3.8}$$

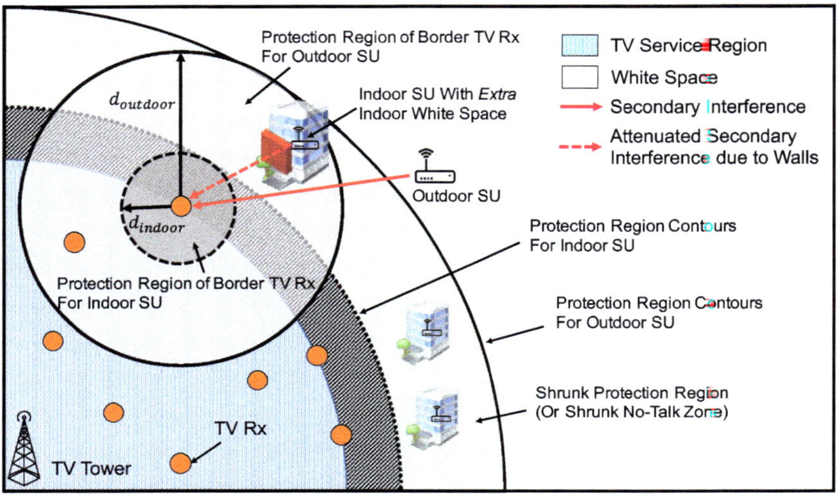


**FIGURE 3.3** Additional white space for indoor SUs. For each border TV Rx, its protection region shrinks from a radius of $d_{outdoor}$ for an outdoor SU to $d_{indoor}$ for an indoor SU, due to built environments (i.e. walls of buildings). Hence, buildings located in the shrunk protection region (in light green) have one additional indoor white space channel.

which means

$$d \geq d_{\min} = L^{-1}\left(E^{SU} - \boldsymbol{L_{wall}} - (E_{Rx}^{TV})_{\min} + \gamma_{\min}\right), \tag{3.9}$$

and

$$E^{SU} \leq (E^{SU})_{\max} = (E_{Rx}^{TV})_{\min} - \gamma_{\min} + L(d) + \boldsymbol{L_{wall}}. \tag{3.10}$$

Fig. 3.3 illustrates the increase in WS availability for an indoor SU. Denote the minimum separation distances for outdoor and indoor WS as $d_{outdoor}$ and $d_{indoor}$, respectively. It is clear that $d_{indoor} < d_{outdoor}$, due to $L_{wall}$ from Eq. (3.9), which is also shown in Fig. 3.3. In other words, an indoor SU located between $d_{indoor}$ and $d_{outdoor}$ from the nearest border TV Rx can see one extra indoor WS channel. Moreover, we can see from Eq. (3.10) that an indoor SU can transmit at a higher power than an outdoor SU.

### 3.3.2 Indoor Grey Space

To use an indoor GS channel, an indoor SU needs to be located within the TV service region. If we only consider outdoor TV Rx, as in the DTV service model, Eq. (3.9) and Eq. (3.10) with $(E_{Rx}^{TV})_{\min}$ replaced by $E_{Rx}^{TV}$ are immediately applicable. Therefore, the same impacts of $L_{wall}$ can be expected: shrunk protection regions (see Fig. 3.4) and increased maximum transmit power for indoor SUs.

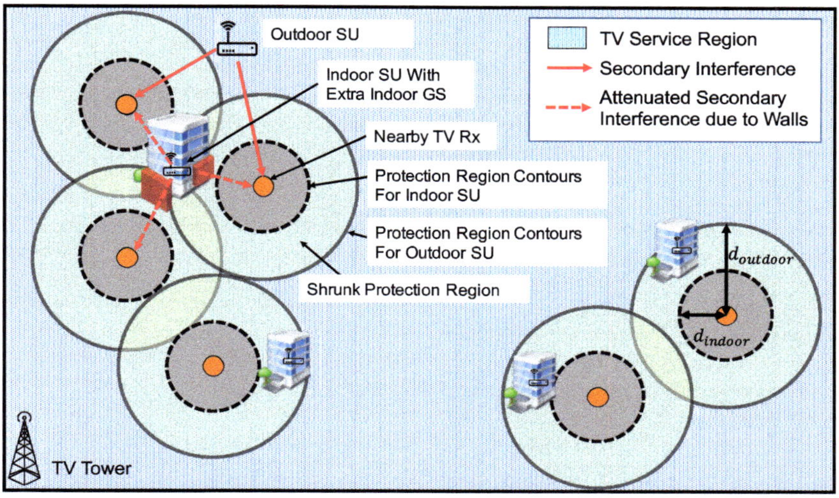


**FIGURE 3.4** Additional grey space for indoor SU. Due to built environments the protection region for a TV Rx shrinks from $d_{outdoor}$ to $d_{indoor}$ in radius when an SU moves from outdoors to indoors. Hence, buildings located in the union of shrunk protection regions (light green rings) but outside the union of protection regions for indoor SUs (grey circles) of all TV Rx's have one additional indoor grey space (assuming no rooftop TV antenna). The presence of rooftop TV antennas results in a smaller set of locations with extra indoor grey space.

However, as we are concerned with buildings (e.g. office and residential) in dense urban areas, we cannot ignore the possibility of installing TV antennas on the roof of a building. It makes much less sense to set up TV antennas inside a building when building administrators can install them on the roof for best TV reception. As a result, such a rooftop TV Rx should be protected from (1) nearby SUs (both indoors and outdoors), and (2) indoor SUs inside the same building. For the first case, previous analysis can be applied to determine the necessary protection regions. In the second case, it is required that

$$[E^{TV} - L_{TV}(d)] - [E^{SU} - L_{indoor}] \geq \gamma_{\min}, \tag{3.11}$$

where $L_{indoor}$ is the overall indoor path loss (due to distance and penetration loss caused by floors and walls) from the indoor SU to the rooftop TV Rx. It implies that

$$L_{indoor} \geq (L_{indoor})_{\min} = E^{SU} - E^{TV} + L_{TV}(d) + \gamma_{\min}, \tag{3.12}$$

and

$$E^{SU} \leq (E^{SU})_{\max} = E^{TV} - L_{TV}(d) + L_{indoor} - \gamma_{\min}. \tag{3.13}$$

Intuitively, if the building is closer to the TV tower or the TV tower has greater transmit power, less $(L_{indoor})_{\min}$ is required and $(E^{SU})_{\max}$ is larger for indoor SUs. On the other hand, for a given building, an SU at an indoor location that provides greater $L_{indoor}$ (e.g. inside a lecture hall on the basement level) is allowed to transmit with greater power without causing harmful interference to the rooftop TV Rx.

In summary, significant penetration loss due to buildings (e.g. walls and floors) effectively weakens secondary interference from an indoor SU, and consequently shrinks protection regions of outdoor TV Rx's. As a result, an SU is able to see more WS and GS indoor than outdoor.

## 3.4 A CASE STUDY IN SAN FRANCISCO

In this section, we first describe the set-up for our case study in San Francisco and our methodology for determining WS and GS from the SU perspective. Then we present path loss models, and compare indoor and outdoor WS/GS availability.

### 3.4.1 Set-Up

In this case study, we assume six-storey buildings with a height of 30 metres which are located at randomly selected locations in San Francisco, as in Fig. 3.5(A). Building footprints are imported from a public dataset [13]. We consider a total of 100+ TV towers within 200 km of San Francisco as in Fig. 3.5(B), whose information is retrieved from UW TV White Space Database SpecObs [14]. Note that the same TV tower may transmit on several channels, which is treated as multiple towers but shown as one on the map.

We consider fixed and portable devices with maximum EIRPs of 36 dBm and 20 dBm, respectively. The antenna height of a rooftop TV antenna is 30 metres, and that of an outdoor TV antenna is 10 metres according to the DTV service model. For simplicity, we assume antenna heights to be 15 metres and 3 metres above ground for fixed and portable secondary devices, respectively, in both outdoor and indoor environments. Since co-channel interference is our primary concern, we use D/U ratios of 34 and 23 dB for analogue and digital TV services, respectively, according to Table 3.A.2.

### 3.4.2 Methodology

In Sections 3.2 and 3.3, we discussed interference protection for WS and GS sharing from the perspective of TV Rx. To determine whether a channel can be claimed as an indoor/outdoor WS/GS channel for a given SU, it is necessary to refine WS/GS from the standpoint of an SU.

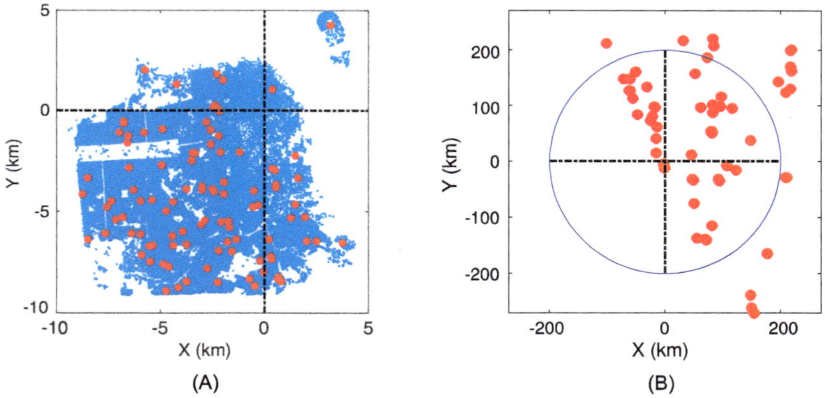


**FIGURE 3.5** (A) Sample topology of 100 random locations in San Fransisco. (B) Locations of nearby TV towers (red dots). All coordinates are in kilometres with respect to the origin at (37.7872, −122.4073). The blue dashed circle in (B) has a radius of 200 km.

Given an SU outside the service region of a TV tower, we assume TV Rx to be everywhere on the borders and consider the nearest border TV Rx as the most vulnerable one. Then we can use Eq. (3.2) through (3.4) or Eq. (3.8) through (3.10) to determine whether a channel is an outdoor or indoor WS channel.

If the SU is within the service region, we check whether a channel is an outdoor and/or indoor GS channel. We consider the following two cases:

- Case 1: both rooftop and nearby outdoor TV Rx's need to be protected.
- Case 2: only nearby outdoor TV Rx need to be protected (no TV Rx on the roof of the same building).

The protection of rooftop TV Rx's is discussed in Section 3.3.2. We now discuss the protection of nearby TV Rx's from the SU perspective. Instead of knowing all TV Rx locations, we assume that we know a circular region of radius $d_{sep}$ around the SU within which no TV Rx is present. Suppose that the SU is located at $d$ from the TV tower. Then we consider the TV Rx with a distance of $d_{sep}$ from the SU and $(d + d_{sep})$ from the TV tower as the most vulnerable one since it receives the strongest interference but the weakest TV signal than the others. For outdoor GS it is then required that

$$[E^{TV} - L_{TV}(d + d_{sep})] - [E^{SU} - L_{SU}(d_{sep})] \geq \gamma_{\min}, \tag{3.14}$$

and for indoor GS that

$$[E^{TV} - L_{TV}(d + d_{sep})] - [E^{SU} - L_{SU}(d_{sep}) - \boldsymbol{L_{wall}}] \geq \gamma_{\min}. \tag{3.15}$$

When $d_{sep}$ is fixed, greater $E^{TV}$, smaller $d$ or larger $L_{wall}$ allows SUs to transmit with greater power. When $d$ is fixed, determining the range of $d_{sep}$ that

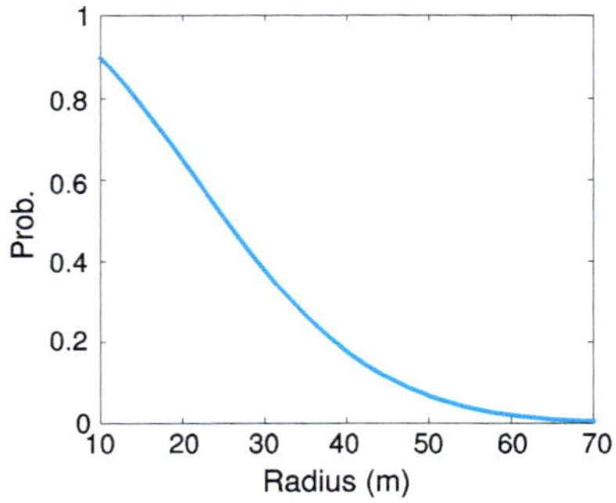


**FIGURE 3.6** Probability of no TV Rx in a circular region of radius $d_{sep}$ in metres.

meets the condition in Eq. (3.14) or (3.15) is not straightforward. Intuitively, when $d_{sep}$ is smaller, secondary interference at the TV Rx is larger and it is harder to meet the above condition, and vice versa. Thus, Eq. (3.14) provides a lower limit on $d_{sep}$, as does Eq. (3.15).

To obtain a reasonable estimate for $d_{sep}$, we make the following assumptions. First, we assume that outdoor TV Rx's (one per house) follow a homogeneous spatial Poisson process [15] with uniform intensity of $\beta > 0$. Denote the no-TV-Rx region (bounded and closed) around an SU Tx as $B \subset \mathbb{R}^2$, and the (random) number of outdoor TV Rx's in $B$ as $N(B)$. Then the probability of $n$ TV Rx's existing in $B$ is given by

$$P(N(B) = n) = \frac{(\beta\lambda_2(B))^n}{n!} e^{-\beta\lambda_2(B)}, \tag{3.16}$$

where $\lambda_2(B)$ is the area of $B$ in km$^2$, and the constant $\beta$ is the expected number of TV Rx's per km$^2$. Hence, the probability of no TV Rx in $B$, which is assumed to be a circular region with a radius of $d_{sep}$ km, is given by

$$P(N(B) = 0) = e^{-\beta\pi d_{sep}^2}. \tag{3.17}$$

Now the task is to find a reasonable approximated value for $\beta$. According to the statistics in [16], San Francisco has a population density of 7022 per km$^2$ and a house density of 2865 per km$^2$. Since 12% of households in the United States on average are still using over-the-air TV broadcasting services [17], it implies an average of 344 houses installed with TV antennas per km$^2$ in San Francisco.

Substituting $\beta = 344$ into Eq. (3.17), we obtain the curve in Fig. 3.6. With $d_{sep} = 20$ m, the probability of no TV Rx is more than 60%, which decreases to less than 20% for $d_{sep} = 40$ m. Hence, we use the range of 20 to 40 metres for $d_{sep}$ in our case study.

### 3.4.3 Path Loss Models

Choosing appropriate outdoor path loss (PL) models is crucial to determining outdoor/indoor WS/GS availability. After careful consideration, we settled on the Longley–Rice (LR) model [18] for outdoor TV signal estimation and the Okumura–Hata model [19] for outdoor secondary interference estimation, both of which were developed from extensive measurements. In addition, we adopted the indoor path loss model recommended by ITU (International Telecommunication Union) [20] for estimating interference from an indoor SU to a rooftop TV Rx.

#### *3.4.3.1 Longley–Rice Irregular Terrain Model*

A computer implementation of the LR model called the irregular terrain model (ITM) is provided by [21]. It takes the following input parameters:

- $f_c$: centre frequency (MHz), between 20 MHz and 20 GHz.
- $d$: distance (km) between the transmitter and the receiver, from 1 km to 2000 km.
- $h_{g1}, h_{g2}$: antenna structural heights (m), from 0.5 m to 3000 m.
- $pol$: horizontal or vertical polarisation.
- $\Delta h$: terrain irregularity parameter or roughness factor (m). This is the main parameter that accounts for terrain elevation in loss calculation.

Note that antenna heights are calculated relative to average terrain height surrounding the TV transmitter. It results in changes in different angles and subsequently non-circular path loss patterns just as the roughness factor $\Delta h$ does.

The output of ITM is a reference attenuation $A_{ref}$ relative to the free space path loss. It is a function of distance $d$ and is computed from the piecewise formula,

$$A_{ref} = \begin{cases} \max\left[0, A_{el} + K_1 d + K_2 \ln\left(\frac{d}{d_{LS}}\right)\right], & d \le d_{LS} \\ A_{ed} + m_d d, & d_{LS} \le d \le d_x \\ A_{es} + m_s d, & d_x \le d \end{cases}, \quad (3.18)$$

where the coefficients $A_{el}$, $K_1$, $K_2$, $A_{ed}$, $m_d$, $A_{es}$, $m_s$ and the distance $d_x$ are calculated using the ITM algorithms. The three intervals are called the line-of-sight, diffraction and scatter regions. All the parameters in Eq. (3.18) are functions of frequency, but this dependency is not straightforward. They also depend on the terrain irregularity parameter, Tx/Rx height, radio climate, polarisation, etc. The total path loss (in dB) is given by

$$L(d) = A_{ref} + 20\log_{10}\left(\frac{4\pi d f}{C}\right), \quad (3.19)$$

where $C$ is the speed of light in vacuum.

In this study, we use the point-to-area mode, where the *point* is the location of a TV tower, and the *area* refers to locations of TV receivers. In this mode, $A_{ref}$ is the median relative attenuation that should be observed on the set of all similar paths during times when the atmospheric conditions correspond to a standard well-mixed atmosphere. More details about LR ITM are available in Appendix B in [11].

### 3.4.3.2 Okumura–Hata Model

This model is a fully empirical prediction method based entirely on extensive measurements for three different environments: urban, suburban and open area [19]. It was also used by the FCC to determine minimum separation distances (see ¶181 in [3]). We chose the model for the urban area for large cities, which is given by

$$L(d) = 69.55 + 26.16\log_{10} f_c - 13.82\log_{10} h_t - a + [44.9 - 6.55\log_{10} h_t] \cdot \log_{10} d, \quad (3.20)$$

and

$$a = \begin{cases} 8.29(\log_{10}(1.54h_r))^2 - 1.1, & \text{if } 150 \leq f_c \leq 300 \\ 3.2(\log_{10}(11.75h_r))^2 - 4.97, & \text{if } 300 < f_c \leq 1500 \end{cases}, \quad (3.21)$$

where $f_c$ is the centre frequency (MHz), $h_t$ is the antenna height (m) of the secondary transmitter, $h_r$ is the antenna height (m) of the TV receiver and $d$ is the distance (km) between the secondary transmitter and the TV receiver.

### 3.4.3.3 Indoor Path Loss Model

This model provides guidance on indoor path loss estimation [20]. It assumes that the pair of transmitter and receiver are located inside the same building, and mainly accounts for the attenuation of the signal through multiple walls and/or multiple floors. In this study, we are interested in the site-general model that requires little path or site information. The basic form is

$$L_{indoor}(d, n) = 20\log_{10} f_c + N\log_{10} d + L_f(n) - 28, \quad (3.22)$$

where $f_c$ is the centre frequency (MHz), $N$ is the distance power loss coefficient, $d > 1$ is the separation distance (m) between the transmitter and receiver and $L_f(n)$ is the floor penetration loss (dB) for $n$ floors between transmitter and receiver ($L_f(0) = 0$ dB). We focus on average path loss and omit shadow fading effects.

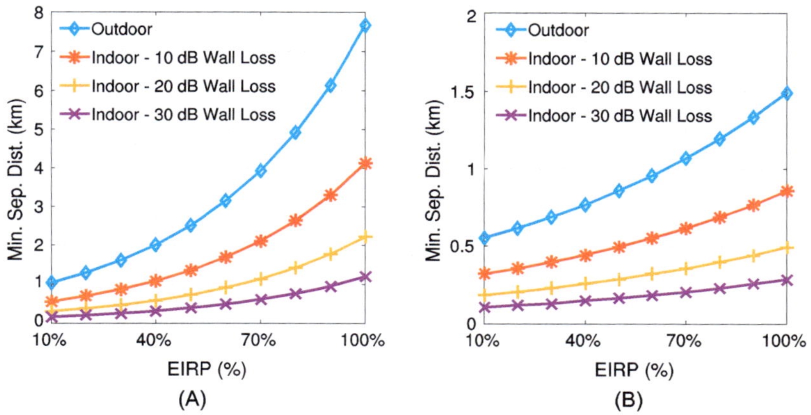


**FIGURE 3.7** Minimum separation distance versus transmit power of secondary devices (in terms of percentage of maximum EIRP). (A) Fixed devices; (B) Portable devices.

## 3.4.4 Indoor Versus Outdoor White Space

In this simulation, we study the impact of SU Tx transmit power $E^{SU}$ and total wall penetration loss $L_{wall}$ on the protection region or minimum separation distance $d_{\min}$ of a border TV Rx. Then we quantify their impact on indoor WS availability in terms of percentage of locations with extra indoor WS.

### *3.4.4.1 Minimum Separation Distance*

According to Eq. (3.3) and (3.9), $d_{\min}$ is a function of secondary user parameters ($E^{SU}$), $L_{wall}$ and FCC rules, that is, $(E_{Rx}^{TV})_{\min}$ and $\gamma_{\min}$, which depend on channel and service type. To understand their impact on $d_{\min}$, we consider a TV tower broadcasting digital TV services on channel 48.

As we can see in Fig. 3.7, $E^{SU}$ has a significant impact on $d_{\min}$ for both outdoor and indoor WS. For instance, if a fixed device transmitting at the maximum allowable EIRP can halve its transmit power, $d_{\min}$ is reduced from 7.7 km to 2.5 km in the outdoor case and from 4.1 km to 1.4 km in the indoor case with 10-dB wall penetration loss.

On the other hand, increasing $L_{wall}$ effectively reduces $d_{\min}$. For fixed and portable devices, if they are located inside buildings when $L_{wall}$ is 10 dB, $d_{\min}$ can be respectively shortened by 46% and 42% on average compared to outdoors. As a result, the reduction in $d_{\min}$ introduces extra indoor WS. For instance, for a full-power fixed device, $d_{\min}$ is 7.7 km for outdoor WS, but contracted to 4.1 km for indoor WS with 10-dB $L_{wall}$. This means that buildings located between 4.1 km and 7.7 km from the protection region contours have an extra indoor WS channel available.

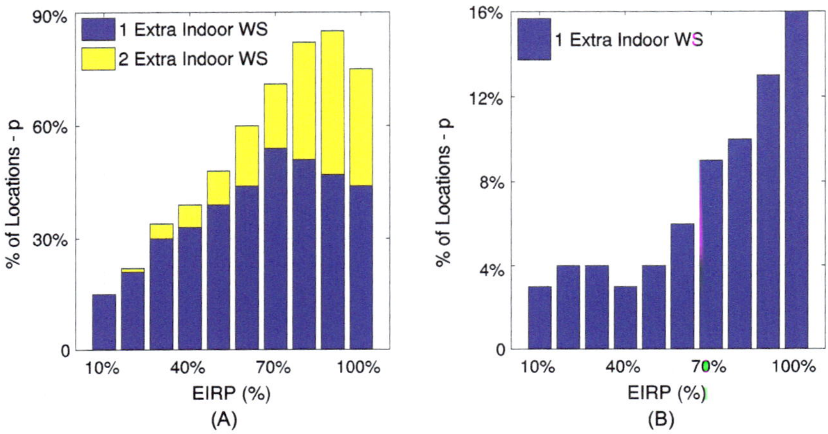


**FIGURE 3.8** Percentage of locations with extra indoor white spaces versus transmit power of secondary devices. (A) Fixed devices; (B) Portable devices.

### *3.4.4.2 Indoor White Space Availability*

In this simulation, we study the impact of $E^{SU}$ on indoor WS availability. For each permissible channel, we determine whether it is an outdoor and/or indoor WS, assuming 15 dB wall penetration loss at each location (100 in total). For each location, we calculate the number of extra indoor WS, and then compute the percentage of locations with extra indoor WS, denoted as $p$.

As shown in Fig. 3.8, $p$ is larger than 70% for fixed devices when the transmit power is greater than or equal to 70% of the maximum EIRP. In contrast, it is less than 20% even in the best case for portable devices. As mentioned previously, only buildings in the contracted no-talk region have an extra indoor WS channel. Therefore, it is necessary for the city of San Francisco to partially overlap with the contracted no-talk region, which means that the city needs to be very close (e.g. several kilometres) to the protection region contours of a TV tower. Since fixed devices have 17 more permissible channels and their contracted no-talk region is much larger than that for portable devices, it makes sense that an extra indoor WS channel is present at many more locations for fixed devices.

Then we set $E^{SU}$ to the maximum and vary $L_{wall}$ from 5 dB to 30 dB with an increment of 5 dB. The result is shown in Fig. 3.9. We observe that $p$ keeps increasing as $L_{wall}$ increases, eventually reaching almost 100% for fixed devices. Note that the percentage of locations with exactly one extra indoor WS increases first and then decreases. As the $L_{wall}$ increases, more locations without any extra indoor WS now have one; as it keeps increasing, more locations with one extra indoor WS now have two, which explains the decrease in the

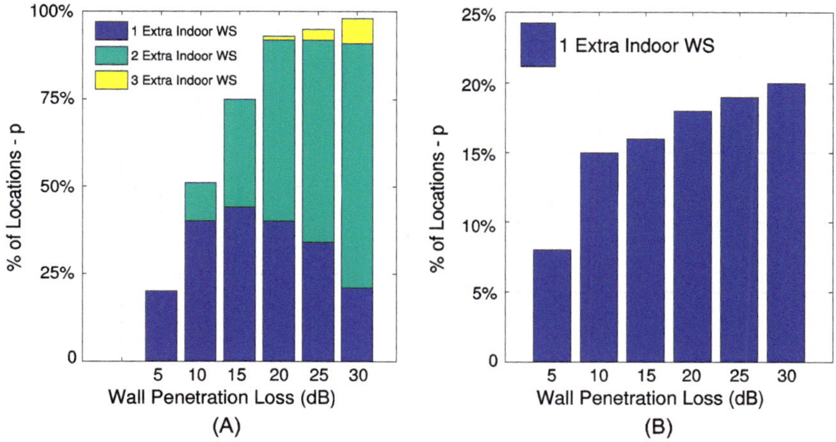


**FIGURE 3.9** Percentage of locations with extra indoor white spaces versus wall penetration loss $L_{wall}$. (A) Fixed devices; (B) Portable devices.

**TABLE 3.1** Average Number of Outdoor White Spaces Versus Transmit Power of Secondary Devices

| EIRP (%) | 10% | 20% | 30% | 40% | 50% | 60% | 70% | 80% | 90% | 100% |
|---|---|---|---|---|---|---|---|---|---|---|
| Fixed | 15.02 | 14.93 | 14.76 | 14.76 | 14.48 | 14.19 | 13.90 | 13.54 | 13.33 | 13.16 |
| Portable | 4.19 | 4.18 | 4.18 | 4.18 | 4.17 | 4.14 | 4.11 | 4.09 | 4.05 | 4.02 |

percentage of locations with one extra indoor WS. We can see a similar effect for $L_{wall}$ for portable devices, but $p$ is much lower for the reasons mentioned previously.

Comparing Figs 3.8 and 3.9, we find that $p$ monotonically increases as $L_{wall}$ increases, but this is not the case for $E^{SU}$. Given a fixed $E^{SU}$, $d_{\min}$ for outdoor WS is fixed, while that for indoor WS keeps decreasing as $L_{wall}$ increases. This means a larger contracted no-talk region, and thus a higher $p$.

On the other hand, given $L_{wall}$, increasing $E^{SU}$ increases $d_{\min}$ and decreases the amounts of outdoor and indoor WS. As shown in Table 3.1, the average number of outdoor white spaces decreases when $E^{SU}$ increases. However, it is difficult to determine the impact on $p$ since it depends on how the contracted no-talk region changes (i.e. increases or decreases in size) and how much it overlaps with the city, but it is certain that the contracted no-talk region will be pushed farther from the protection contours as transmit power increases. This may explain the decrease in $p$ when $E^{SU}$ decreases from 90% to 100% in Fig. 3.8(A).

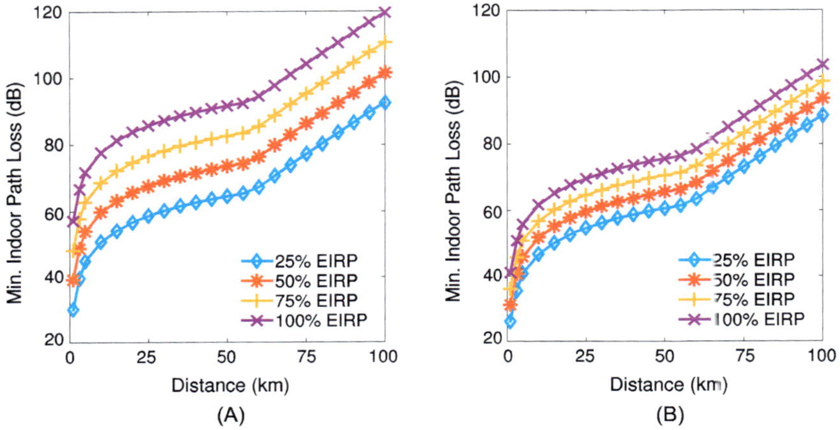


**FIGURE 3.10** Minimum indoor path loss required for rooftop TV Rx protection versus the distance between the building and the TV tower for different transmit power of secondary devices. (A) Fixed devices; (B) Portable devices.

## 3.4.5 Indoor Versus Outdoor Grey Space

In this simulation, we first explore conditions on indoor path loss for rooftop TV antenna protection, and study the impact of building locations and wall penetration loss $L_{wall}$ on the minimum separation distance $(d_{sep})_{\min}$ (within which no TV Rx is present). Then we study the impact of $E^{SU}$ and $L_{wall}$ on indoor GS availability.

### 3.4.5.1 Conditions for Rooftop TV Antenna Protection

As mentioned in Section 3.4.2, we consider protecting the rooftop TV Rx from the SU inside the same building in Case 1. As an example, we consider the 1000 KWatt TV tower at (38.27, −121.49) that broadcasts digital services on CH 48 at 755 MHz, and a building located due east of the tower at a distance $d$.

As illustrated in Fig. 3.10, when the building is farther away from the TV tower, the TV signal is weaker and the required $(L_{indoor})_{\min}$ is greater. For a fixed $d$, the minimum $L_{indoor}$ increases linearly as $E^{SU}$ increases. If we consider an indoor SU Tx on the fourth floor with a distance of 15 metres from the roof and a separation of three floors, then the indoor path loss is

$$L(15,3) = 20\log(755) + N\log_{10}(15) + L_f(3) - 28 = 92.4 \text{ (dB)},$$

where $N = 33$ and $L_f(3) = 24$ dB as suggested by [20]. It means that the building is at most 55 km and 80 km from the TV tower for full-power indoor fixed and portable devices, respectively. With smaller transmit power and larger in-

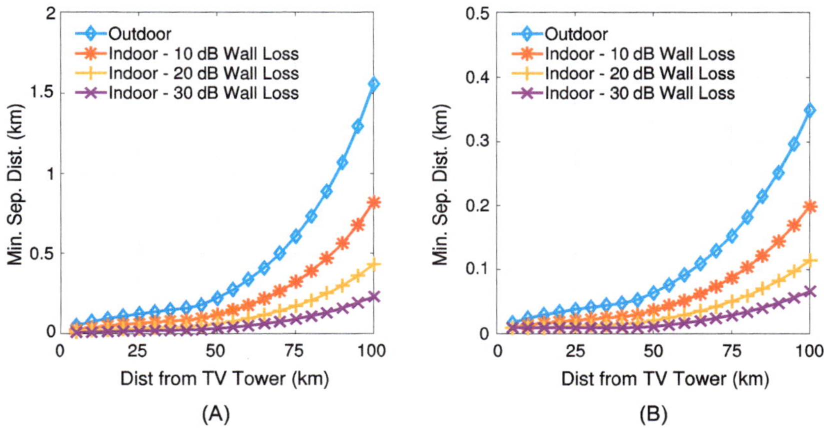


**FIGURE 3.11** Minimum separation distance versus the distance between the building and the TV tower. (A) Fixed devices with maximum EIRP; (B) Portable devices with maximum EIRP.

door path loss, the maximum allowable distance between the building and the TV tower can be larger.

### *3.4.5.2 Minimum Separation Distance*

According to Eq. (3.14) and (3.15), $(d_{sep})_{\min}$ for GS depends on $E^{SU}$, $L_{wall}$, $\gamma_{\min}$ (specified by FCC) and $d$ (the distance between the building and the TV tower). Here, we consider the same TV tower as in the previous section, and assume the maximum EIRP for both fixed and portable devices.

As shown in Fig. 3.11, when the SU is closer to the TV tower, TV signal strength is stronger and nearby TV Rx's can operate closer to the SU. For instance, to claim an outdoor GS channel at a distance of 50 km from the TV tower, a fixed device needs to ensure no presence of TV Rx within approximately 220 metres. In contrast, $(d_{sep})_{\min}$ can be shortened to 120 metres, if the device is 25 km from the TV tower.

We also observe that $(d_{sep})_{\min}$ decreases as $L_{wall}$ increases since it weakens the secondary interference. For a given $(d_{sep})_{\min}$, a larger $L_{wall}$ introduces additional indoor GS. For instance, when $(d_{sep})_{\min} = 0.5$ km, the channel is not an outdoor GS channel for a fixed device located at 75 km from the TV tower, but it is an indoor GS channel if the device is in a building that provides at least 10-dB $L_{wall}$.

### *3.4.5.3 Indoor Grey Space Availability*

In this simulation, we consider the two cases for indoor GS, as discussed in Section 3.4.2. For Case 1, we assume an indoor SU on the fourth floor with a distance of 15 metres and a separation of three floors from the roof. We also

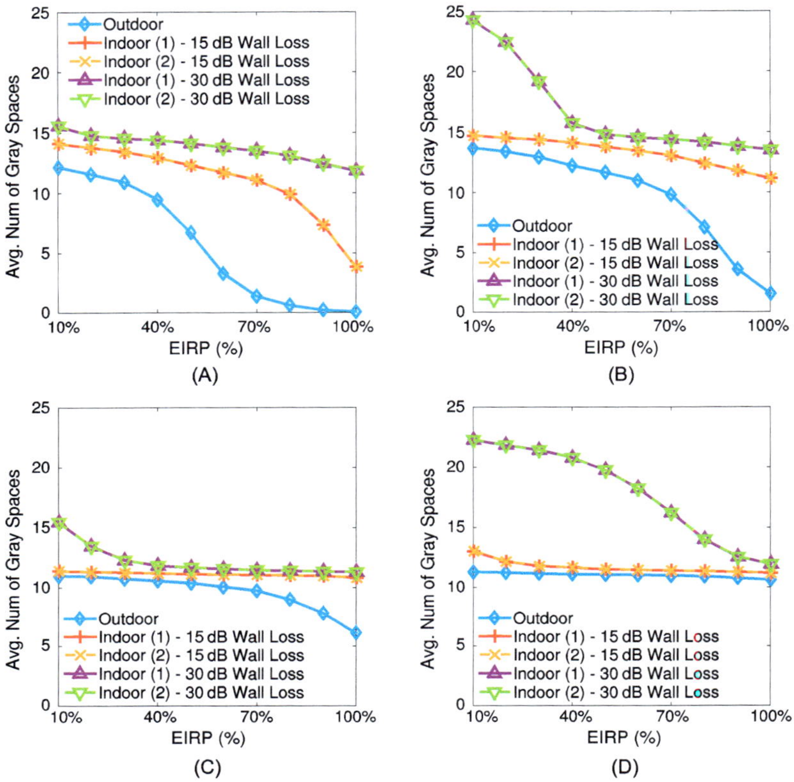


**FIGURE 3.12** Average number of grey spaces versus transmit power of secondary devices and wall penetration loss. Indoor (1)/(2) in the legend means Case 1/2 for indoor grey space. (A) Fixed devices, $d_{sep} = 20$ m; (B) Fixed devices, $d_{sep} = 40$ m; (C) Portable devices, $d_{sep} = 20$ m; (D) Portable devices, $d_{sep} = 40$ m.

assume $N = 33$ and $L_f(3) = 24$ dB for the indoor path loss model. For each channel at each location, we determine whether it is an outdoor and/or indoor GS, and then compute the number of GS averaged over 100 locations for different settings. We set $d_{sep}$ to 20 m or 40 m. The results are provided in Fig. 3.12.

First, we observe that the average number of indoor GS in Case 1 is the same as in Case 2, which means that the rooftop TV Rx protection in the above setting is not a limiting factor for indoor GS availability. Second, a larger $d_{sep}$ leads to an increase in the average number of outdoor and indoor GS channels for both fixed and portable devices. We also see similar effects of decreasing $E^{SU}$ and increasing $L_{wall}$. Third, there are more GS channels indoors than outdoors. Given $L_{wall}$ and $d_{sep}$, the amount of extra indoor GS does not necessarily increase monotonically as $E^{SU}$ increases because more outdoor GS appear as well. This is the case, however, when we increase $L_{wall}$, provided $E^{SU}$ and $d_{sep}$ are fixed.

Overall, in a major city like San Francisco with a dense population and houses, the area without any TV Rx presence is very small for a fixed device. Hence, to get more GS channels (either outdoors or indoors), it will have to lower its transmit power or operate in indoor environments that provide significant penetration loss (e.g. in an auditorium room on the basement level). Portable devices are less likely to interfere with nearby TV Rx's, due to smaller transmit power, but wall penetration loss can sometimes still lead to a significant gain in GS availability.

## 3.5 CONCLUSION

The global transition from analogue to digital OTA TV broadcasting has freed up a large portion of spectrum, and the FCC has announced a set of rules to enable spectrum sharing between incumbents and TVBDs. In this study, we revisit the interference protection criteria for outdoor WS in accordance with the FCC ruling, and extended them to outdoor GS and also to indoor WS/GS. We studied the impact of built environments, and performed a US case study to compare outdoor and indoor WS/GS availability. Our results show that there are more WS channels indoors than outdoors, due to the attention effects of walls on secondary interference. Further, there are sufficient GS opportunities outdoors, and many more indoors.

## ACKNOWLEDGEMENT

This work was supported in part by NSF EARS Award 1443923.

## APPENDIX 3.A FCC OPERATION RULES

**TABLE 3.A.1** Thresholds (in dBu) for TV Station Protection Contours

| Type of Station | Protection contour | | |
|---|---|---|---|
| | *Channel* | *Contour (dBu)* | *Propagation curve* |
| Analogue: Class A, LPTV, translator and booster | Low VHF (2–6) | 47 | F(50, 50) |
| | High VHF (7–13) | 56 | F(50, 50) |
| | UHF (14–51) | 64 | F(50, 50) |
| Digital: Full service TV, Class A TV, LPTV, translator and booster | Low VHF (2–6) | 28 | F(50, 50) |
| | High VHF (7–13) | 36 | F(50, 50) |
| | UHF (14–51) | 41 | F(50, 90) |

**TABLE 3.A.2 Desired-to-Undesired Signal Ratio Defined by FCC for Maximum Tolerable Interference in Various TV Applications. We Primarily Consider Co-Channel Interference in This Study**

| Type of Station | Protection ratios | |
|---|---|---|
| | *Channel separation* | *D/U ratio (dB)* |
| Analogue TV, Class A, LPTV, translator and booster | Co-channel | 34 |
| | Upper adjacent | −17 |
| | Lower adjacent | −14 |
| Digital TV and Class A | Co-channel | 23 |
| | Upper adjacent | −26 |
| | Lower adjacent | −28 |

## REFERENCES


[1] J.M. Peha, The digital TV transition: a chance to enhance public safety and improve spectrum auctions, IEEE Communications Magazine 44 (6) (2005) 22.

[2] M. Murroni, S. Scalise, A. Vanelli-Coralli, S. Kim, R. Briskman, Convergence of digital TV systems and services, International Journal of Digital Multimedia Broadcasting 2009 (2009) 817636.

[3] In the matter of unlicensed operation in the TV broadcast bands: second report and order and memorandum opinion and order (FCC Std. 08-260).

[4] In the matter of unlicensed operation in the TV broadcast bands: second memorandum opinion and order (FCC Std. 10-174).

[5] T. Yucek, H. Arslan, A survey of spectrum sensing algorithms for cognitive radio applications, IEEE Communications Surveys and Tutorials 11 (1) (2009) 116–130.

[6] V. Gonçalves, S. Pollin, The value of sensing for tv white spaces, in: 2011 IEEE Symposium on New Frontiers in Dynamic Spectrum Access Networks (DySPAN), IEEE, 2011, pp. 231–241.

[7] H. Bezabih, B. Ellingsaeter, J. Noll, T. Maseng, Digital broadcasting: increasing the available white space spectrum using TV receiver information, IEEE Vehicular Technology Magazine 1 (7) (2012) 24–30.

[8] J. Huang, R.A. Berry, M.L. Honig, Spectrum sharing with distributed interference compensation, in: First IEEE International Symposium on New Frontiers in Dynamic Spectrum Access Networks, DySPAN 2005, IEEE, 2005, pp. 88–93.

[9] K. Harrison, S.M. Mishra, A. Sahai, How much white-space capacity is there? in: 2010 IEEE Symposium on New Frontiers in Dynamic Spectrum, IEEE, 2010, pp. 1–10.

[10] D. Makris, G. Gardikis, A. Kourtis, Quantifying TV white space capacity: a geolocation-based approach, IEEE Communications Magazine 50 (9) (2012) 145.

[11] F. Hessar, S. Roy, Capacity considerations for secondary networks in TV white space, IEEE Transactions on Mobile Computing 14 (9) (2015) 1780–1793.

[12] Digital Mobile Radio Towards Future Generation Systems: COST 231 final report, Tech. rep., European COST Action 31, 1999.

[13] Building footprints of the city and county of San Francisco as of June 2011, https://data.sfgov.org/Housing-and-Buildings/Building-Footprints/6p33-sdua (Accessed 27 October 2016).

[14] F. Hessar, S. Roy, Cloud based simulation engine for TVWS, http://specobs.ee.washington.edu/.

[15] A. Baddeley, I. Bárány, R. Schneider, Spatial point processes and their applications, in: Stochastic Geometry: Lectures Given at the CIME Summer School Held in Martina, Franca, Italy, 2007, pp. 1–75.
[16] San Francisco, CA houses and residents, http://www.city-data.com/housing/houses-San-Francisco-California.html (Accessed 29 November 2016).
[17] The total Nielsen audience report: Q2 2016, September 2016, http://www.nielsen.com/us/en/insights/reports/2016/the-nielsen-total-audience-report-q2-2016.html.
[18] Longley–Rice Methodology for Evaluating TV Coverage and Interference, Tech. Rep. OET Bulletin No. 69, Federal Communications Commission, February 2004.
[19] A. Aragon-Zavala, Antennas and Propagation for Wireless Communication Systems, John Wiley & Sons, 2008.
[20] Propagation data and prediction methods for the planning of indoor radiocommunication systems and radio local area networks in the frequency range 300 MHz to 100 GHz (Recommendation ITU-R, P.1238-8).
[21] G.A. Hufford, The Its Irregular Terrain Model, Version 1.2.2, The Algorithm, Tech. rep., August 2002, http://flattop.its.bldrdoc.gov/itm.html.

Chapter 4

# TV White Space Developments in the European Union

**Jussi Kahtava**
*Allied Spectrum Associates Ltd, Fleet, United Kingdom*

## 4.1 INTRODUCTION

TV White Spaces (TVWS) emerged in the USA as an opportunistic way of using TV broadcast frequencies where the unused frequencies left by frequency reuse by broadcast towers (known as white spaces) allowed emission of radio signals by lower power devices. In the United Kingdom, Ofcom adopted this concept in the latter part of the previous decade as one way of enhancing flexible use of spectrum. Ofcom proposed a number of different changes to the Federal Communications Commission (FCC) rule making, and after a round of consultations and a 12-month pilot, enacted new regulations legalising access of TVWS devices on UHF bands in 2015. In continental Europe, for reasons such as the difficulty in identifying the existing unlicensed PMSE use and general concerns regarding the novelty of opportunistic access, TVWS regulation has not advanced to the extent it has in the United Kingdom. On the other hand, the technique of geolocation databases was adopted from TVWS and is being deployed in other methods of shared use of spectrum. One key example in Europe is Licensed Shared Access on the 2300–2400 MHz band. Standardisation in Europe takes place under the aegis of the European Telecommunications Institute (ETSI) [1] which has developed both TVWS radio standards (in ETSI BRAN) and geolocation database standards (in ETSI RRS) under the Reconfigurable Radio Systems mandate issued by the European Commission.

## 4.2 THE STATUS OF UHF BAND IN EUROPE

The use of UHF band in Europe was recently reviewed by the Lamy commission, set up by the European Commission. This was part of the ongoing deliberation on the future of terrestrial broadcast in the European Union member states. Whilst some countries such as the United Kingdom continue to see benefit in Over The Air (OTA) broadcasts on UHF bands, some other countries

**TV White Space Communications and Networks. http://dx.doi.org/10.1016/B978-0-08-100611-5.00006-0**

are ready to give up on OTA television. Due to the high-tower high-power nature of broadcast TV, the border coordination issues are of crucial importance and a harmonised approach to releasing TV spectrum bands for other purposes would be beneficial. It was this need for a common approach that underlined the elaborations by the Lamy commission. The 20-25-30 report issued by the Lamy commission together with the outcome of the World Radio Conference 2015 (which issued a moratorium on broadcast assignments until WRC 2023) has stabilised the situation of broadcast services in Europe for the next decade to come. In principle, the remaining broadcast spectrum that is left between 470 MHz and 694 MHz has room to share with TVWS devices.

It must be remembered that traditionally the programme making and special events (PMSE) community has been the 'opportunistic sharer' on TV bands [2]. For a number of years the PMSE community has raised concerns regarding the diminishing TV spectrum after consecutive digital dividends at 800 MHz and now on 700 MHz. Packing TV channels into narrower spaces also leaves less room for other uses such as wireless microphones that have exploited the white spaces. One example given is that in some venues the number of available 8 MHz TV channels may go down to only four, which leaves too little room for PMSE use at larger events at such a place (Alan March at UK Spectrum Policy Forum, March 2016).

The policy makers in Europe have realised the challenges facing wireless microphones in the future and alternative schemes have been considered. One proposed option is the 960–1164 MHz band, which according to Ofcom studies would be usable for radio microphones. Whether the PMSE community will move to using that band is not yet clear, but what is clear is that the white space available for either PMSE or white space devices in the post-700 MHz world has become narrower.

Beyond the United Kingdom, the concept of licence-exempt sharing on the white spaces of this band has not really gained traction in the European countries. One important reason for the lack of initiative is the more complicated sharing scenarios with PMSE. Unlike in the United Kingdom, other European countries do not have existing databases of licensed microphones; actually, it is very difficult to tell where and how PMSE is being used on the UHF band in most European countries. Some solutions have been proposed, such as the PMSE web registration (by Fairspectrum), but in general, the administrations have not been proactively promoting the access of TVWS devices to the TV bands. This lack of existing incumbent information in an easily digitised format makes it difficult to establish the geolocation database on incumbent protection from sharing use. The combination of the added effort to even have a database on the incumbent use and the uncertainty about the amount of TVWS left after a 700 MHz broadcast is repacked with other broadcast use on the 490–694 MHz frequency band makes the emergence of TVWS applications in continental Europe a remote possibility.

The Finnish communications regulatory authority, FICORA, has been one of the more forward-looking entities in continental Europe. They have proactively supported trials in TVWS in particular, and have shared use of spectrum in general during this decade. In fact, Finland was the first country in Europe where a geolocation database license was issued: in 2012 trials were conducted in the city of Turku. Despite the early interest, little has happened in Finland with regards to commercial operations of TVWS on the TV bands. Recent discussions with Ficora in March 2016 did not indicate that commercial operations would be starting anytime soon in Finland.

## 4.3 STANDARDISATION FRAMEWORK IN EUROPE

European standards development for Information and Communication Technologies (ICT) takes place under a multi-faceted regulatory environment where the policy making, regulatory and standards developing bodies interact. ETSI is one of the three official standards development organisations which develop standards under the aegis of European Commission policies. The other two are CEN (Comité Européen de Normalisation) and CENELEC (Comité Européen de Normalisation Électrotechnique). Fig. 4.1 shows the three-pronged framework where the European Commission issues mandates to Electronic Communications Committee (ECC) (on spectrum management) and ETSI, CEN and CENELEC (on standards development). The ECC issues CEPT (European Conference of Postal and Telecommunications Administrations) reports as a response to the mandate, whereas ETSI, for example, develops harmonised standards. Finalised harmonised standards are published in the Official Journal of European Commission. ETSI and ECC have a Memorandum of Understanding in place under which the two bodies exchange technical requests and information.

ETSI explains that the standards provide the following benefits:

- Safety and reliability – Adherence to standards helps ensure safety, reliability and environmental care. As a result, users perceive standardised products and services as more dependable, which in turn raises user confidence, increasing sales and the adoption of new technologies.
- Support of government policies and legislation – Standards are frequently referenced by regulators and legislators to protect user and business interests, and to support government policies. Standards play a central role in the European Union's policy for a Single Market.
- Interoperability – The ability of devices to work together relies on products and services which comply with standards.

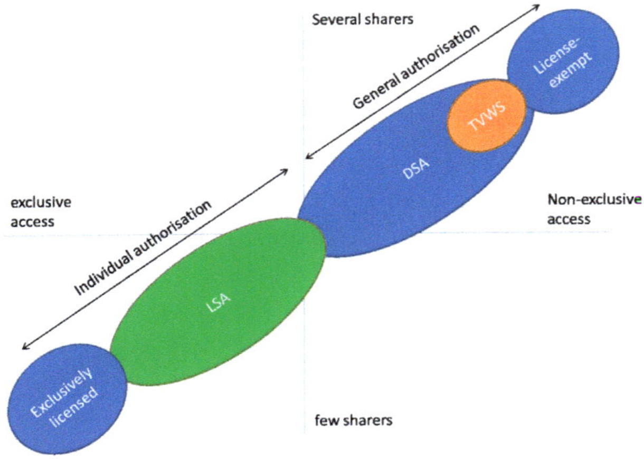


**FIGURE 4.1** The continuum of spectrum authorisation mechanisms.

- Business benefits – Standardisation provides a solid foundation upon which to develop new technologies and to enhance existing practices. Specifically standards can open up market access, can provide economies of scale, can encourage innovation and can increase awareness of technical developments and initiatives.
- Consumer choice – Standards provide the foundation for new features and options, thus contributing to the enhancement of our daily lives. Mass production based on standards provides a greater variety of accessible products to consumers.

When spectrum is needed for a new application, ETSI develops what is known as a System Reference Document (SRDoc), which outlines the requirements and objectives for the candidate standard and its spectrum needs. This SR-Doc is submitted to the ECC where the Frequency Management Working Group (WG FM) considers how the new requirements may be fit within the existing spectrum management framework of CEPT member states. The Spectrum Engineering Working Group (WG SE) eventually gets involved and pursues technical studies towards practical spectrum harmonisation targets in the band under consideration. As an outcome of its deliberations the ECC issues a decision or a recommendation. The decisions of the ECC are not binding on CEPT member states and the national regulatory authorities have flexibility in deciding whether they take the spectrum harmonisations mechanisms into use for the new radio standard. In some cases the spectrum harmonisation framework is extended into a EC decision which is binding on European Union member states. The Radio Spectrum Committee of the European Commission is involved in this process.

An important aspect of the harmonised standards today is the sharing of frequency bands between different applications or services. Sharing as such has taken place for a long time; good examples are the use of PMSE on TV bands or the common use of C-band for Fixed Service and Fixed Satellite Service. However, TVWS is a good indication of the direction where sharing is developing, and all the potential scenarios of mitigation techniques and sharing procedures are not likely to be written into spectrum regulatory documents. The Harmonised European Norms that ETSI produces will include more detailed techniques, which in turn streamlines and eases the regulatory environment. Thus, harmonised standards become ever more important and do not lose their role in efficient market entry and equipment circulation even during the oncoming cognitive radio era.

The smooth introduction and circulation of equipment to the European market is the third prong of the regulatory landscape in the ICT field. In the European Union, the new Radio Equipment Directive (RED) – which replaced the earlier R&TTE directive – came into force in 2016. This framework has been crucial in reducing the need for country or operator specific type approval processes when a manufacturer can declare that its products are in full conformance of the stipulations of the directive. The conformance may be assessed by the Notified Body or by ensuring that the product fulfils the requirements of the relevant Harmonised European Norm.

## 4.4 DEVELOPMENTS WITH GEOLOCATION DATABASES

The fundamental tool for making opportunistic sharing work in TVWS is the geolocation database. Starting early on, Europe and the United States took different approaches to the databases and how they communicate. In the United States, where all the geolocation databases are expected to have exactly the same information available for a device that is querying from the same coordinates, it did not make much sense for the FCC to influence inter-database connections. These were left for a group of database providers to agree on among themselves. In Europe, on the other hand, the national regulatory authorities never had much appetite for a pan-European super-database which would serve devices in multiple member states. A much more likely outcome was for independent geolocation databases in each country with the consequent border coordination issues that always emerge when spectrum is deployed on neighbouring constituencies in close proximity to one another. With this eventuality in mind, standards development in Europe focused on open standardised interfaces not only between TVWS devices and the geolocation database, but also between geolocation databases.

In November 2012 the European Commission issued mandate M/512 to European standards developing organisations (ETSI, CEN, CENELEC) regarding reconfigurable radio systems. This was a part of a larger scope of shared use that at that time covered the aspects of both TVWS and the emergent Licensed Shared Access framework. The author remembers attending a workshop at the JRC Ispra site in 2011 where the guidelines of the mandate were hammered out. The intent was to encourage dialogue between the defence, civil security and commercial sides on sharing spectrum in order to find potential areas of synergy (objective C of the mandate). However, to date there is not much evidence of anything having taken place under objective C. Under the aegis of the mandate M/512, ETSI has developed a number of standards (European Norms) towards commercial TVWS use:

- EN 303 143 (TS 103 143): System Architecture for information exchange between different GLDBs;
- EN 303 144: Cognitive radio systems and geolocation databases;
- EN 303 145 (TS 103 145): System architecture for TVWS use;
- EN 303 387: Interface between CRS and SC.

EN 303 143 specifies the architecture for information exchange between geolocation databases. As mentioned earlier, information exchange between national databases is important in Europe where there are many border coordination issues to be considered. Without a European Norm the risk of non-compatible databases entering the market is too large, and automated coordination becomes very difficult.

## 4.5 EXTENSION OF THE TVWS CONCEPT TO OTHER BANDS

The work that did take place on geolocation databases in North America and in ETSI and CEPT in Europe has born fruit in the wider domain of shared use of spectrum. The Licensed Shared Access (LSA) framework that is emerging on the 2300–2400 MHz band in Europe relies on the use of geolocation databases in the form of LSA repositories and LSA controllers. The principle between the database operation in TVWS and LSA is, in general terms, the same:

- Collection of incumbent information;
- Provision of incumbent data to the calculation engine;
- Translation of incumbent information and regulatory constraints into operational parameters;
- Provision of operational parameters to the radio network manager of the sharing user.

The operation of a geolocation database is 'agnostic' to which actual frequency band it is serving. Naturally, the incumbent use is different from one band to another, and this is reflected in what information is collected, and how often the database has to be queried by sharing users.

Interestingly, when we look at the different sharing schemes on a chart where they are shown in terms of exclusivity vs. number of rights holders, there is a continuum from fully licence-exempt use through various sharing mechanisms up to fully exclusive licence (see Fig. 4.1).

Individual authorisations: Individual authorisations given to one or a few users selected by the regulator, approach ranging from simple written license only in a static case to license plus technically facilitated varying secondary spectrum use. Suitable for mobile operators, if average capacity is predictable and Quality of Service (QoS) is facilitated. In its simplest form, it is very similar to current exclusive cellular licensing.

TVWS: No limitation on users or their amount. First come, first served, technically means that sensing or database may be used to indicate free portions of band and announce possible technical conditions on the usage. In this form of license-exempt use where no QoS is facilitated through the authorisations.

LSA: Users of LSA may be a group selected by a regulator, or it may be open to anyone. Technically means that the database that authorises usage of a certain part of spectrum for a certain time, based on usage requests. Authorisations facilitate Coexistence and QoS. Number of users may need to be limited in case use requests exceed the capacity.

Of particular interest in Europe at the moment is the emergence of C-band for land mobile communications. The band of 3400–3600 MHz had already been identified for IMT by CEPT in 2007, and more global scale was achieved at WRC 2015. This band is fairly clear of previous use (satellites) and it is expected to be released for mobile network operators in due course. However, the band of 3600–3800 MHz – despite being promoted by the CEPT – did not get support from the majority of the ITU member states at the World Radio Conference. Nevertheless, the CEPT is pursuing the assignment of this band for IMT, but it is still being used by satellite services. One proposal is to establish a common European LSA repository at the European Communications Office in Denmark, which would tap into the incumbent spectrum use information on that band and provide it to the radio resource managers of the sharing users. The principle of the original TVWS geolocation database is again revisited in that concept. (See Fig. 4.2.)

In particular, the consultation that UK Ofcom issued on 14 April 2016 [3] is noticeable in its idea that a three-tier assignment akin to the US Citizens Broadband Radio Service (CBRS) would be adopted on the 3800–4200 MHz band in the United Kingdom. This band is allocated in CEPT for Fixed Links and Fixed Satellite Services (space-to-earth) on primary basis and for land mobile on secondary basis. It also has assignment for UK Broadband. At the time of writing the proposal was under consultation and the final concept may change. Ofcom considered that the approach authorised for TVWS-type devices at television bands could be reused at the lowest tier of the three:

**Tier 1:** Existing licence classes

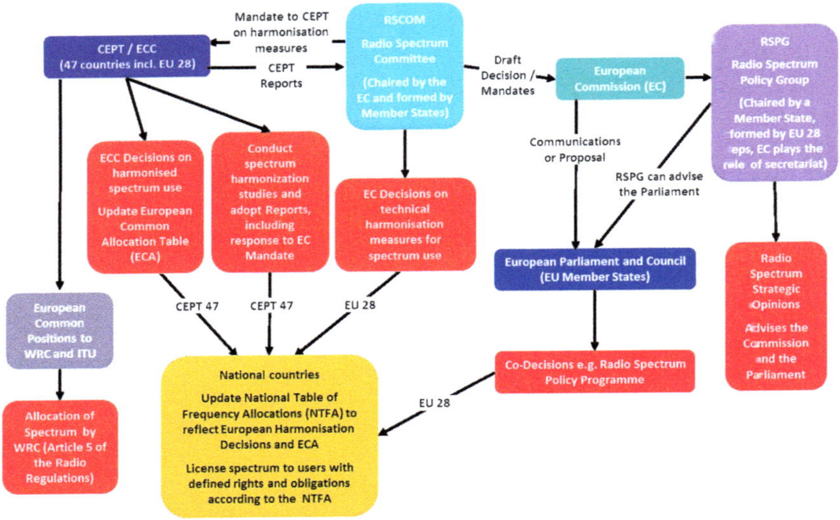


**FIGURE 4.2** Radio spectrum management process in Europe.

**Tier 2:** New geographic licence layer
**Tier 3:** New opportunistic access layer.

Existing licences refer to the primary assignments for Fixed Satellite Service (FSS) earth stations and fixed links situated mainly in the London area and north-east Scotland. Some growth is seen in the utilisation of the band for these incumbent services and spectrum is assigned on a first come, first served basis by the regulator. It is considered that the fixed nature of the incumbent use leaves plenty of spatial domain free with no use and any given fixed link installation only uses some of the frequency bands available for FS.

The common denominator between the FCC and Ofcom schemes is that the third tier is best effort, and is obliged to avoid interfering with the tiers above it. The use of geolocation databases is envisioned to assist the tier 3 networks and devices to avoid such interference. Thus, similarities to TVWS are obvious and the key difference is the nature and type of incumbents and other protected entities.

All in all, what we can see in the shared use of spectrum in continental Europe is that the TVWS concept of geolocation-based sharing is going ahead, albeit not at the original band of intent of UHF broadcast services.

The extension of the TVWS approach to sharing on a licence-exempt basis can be understood as Dynamic Spectrum Access (DSA). In the context of the European regulatory developments, it is this more generic aspect of opportunistic access to radio spectrum resources that may finally open the gates to what was initially intended with TVWS technology. DSA, however, has different challenges from TVWS. Whereas it was difficult to have administrations agree on opportunistic shared use of UHF bands, the TV band itself is globally assigned to similar use, and TVWS devices in principle had a very wide basis for establishing scale benefits in production. More general DSA, on the other hand, does not today have an obvious target band that would be similarly harmonised and aligned in global use.

The latest development in shared use of spectrum is the adoption of Long Term Evolution (LTE) in licence-exempt bands. This started as Licence-Assisted Access (LAA) in 3GPP Rel-13 where the downlink from a base station may have a component carrier on licence-exempt band, whilst uplink is transmitted on operator's licensed band. Rel'14 extends this concept (eLAA) into also having an uplink carrier in licence-exempt band. In both cases, the system is very much under the control of an operator network management, and the key question is coexistence, for example with existing Wi-Fi access points.

An interesting extension of the LAA/eLAA is a concept promoted by MulteFire Alliance (MFA) [4] where an LTE system is completely independently deployed on licence-exempt bands. One particular target for this concept is at the US CBRS at the 3550–3700 MHz band where LTE devices would operate at the GAA layer. Outside of the US CBRS, the MFA has also expressed interest in deploying this technology at other bands, such as sub-1 GHz for Internet of Things (IoT) applications. The Spectrum Access System (SAS) to be used on the US CBRS band will naturally extend for example to the 902–928 MHz licence-exempt band in the US market. In Europe, a robust coexistence mechanism with Wi-Fi on ISM bands or with other licence-exempt devices on the 868–870 MHz band will again benefit from geolocation database schemes.

When we think back to the European Commission radio spectrum review that took place earlier this decade, it was evident that many incumbents or administrations were reluctant to disclose how and where some spectrum bands were being used. This introduced another way of talking about white space: empty blocks in the frequency tables where little information was available on the intensity and type of spectrum use. Ideally, such swathes of spectrum across a region where a band is a potential target for sharing provides a good ground for DSA schemes.

All devices, whether they are expensive smart phones or cheap wearables, benefit from harmonised spectrum which makes it much more cost-efficient to produce equipment to the consumer market. In the European Union, harmonised standards are the preferred route for easy market access to manufacturers. Thus, even in the context of DSA where devices are expected to use open spectrum when and where it is available, the same practicalities face antenna and radio front-end design as in any mass market radio device. Using a band that is available across a region in a reasonably harmonised fashion, albeit at different incumbent occupancy ratio and power limits, etc., would be much more attractive for DSA than the concept of generic hopping here and there in radio frequencies.

## 4.6 CONCLUSIONS

Although it can be argued that the business models originally envisioned for TV White Spaces have not quite emerged as quickly as was thought, the impact of the flexible spectrum use concept spearheaded by TVWS is alive and well. We have seen the geolocation database develop from a repository of locations and time instants unavailable for opportunistic access devices (to protect incumbent use) into a multi-faceted tool which will enable the harmonious coexistence of several types of radio technologies in bands extending from UHF to 2300–2400 MHz, to C-band and beyond. The original challenge to the established mindset in spectrum regulation has enabled the national regulatory bodies and industry to have a more open discussion on the possibilities available to increase the fair and efficient use of radio spectrum. Even though the actual regulation on TVWS use on UHF bands has not really happened in continental Europe, we can rest assured that the fruits of the concept emerge in LSA regulation and forthcoming C-band sharing.

## REFERENCES

[1] ETSI, Why we need standards, http://www.etsi.org/standards/why-we-need-standards.
[2] Ofcom statement, New spectrum for audio PMSE, 10th March 2016.
[3] Ofcom consultation, 3.8 GHz to 4.2 GHz band: opportunities for innovation, 14th April 2016.
[4] MulteFire Alliance Rel 1.0 White paper, MFA website 17th January 2017.

Chapter 5

# TV White Space Technology for Affordable Internet Connectivity

**Meghna Khaturia, Sarbani Banerjee Belur and Abhay Karandikar**
*Indian Institute of Technology Bombay, Mumbai, India*

## 5.1 INTRODUCTION

In the modern digital age, the 'Right to Ubiquitous Connectivity' is a basic human right of individuals irrespective of where they live, their socio-economic status and gender. Internet has played a major role in the connectivity revolution by allowing the maximum number of people to enjoy its benefits. In spite of this spectacular growth in internet usage, a staggering 47% of the global population is still unconnected. Furthermore, the percentage of the unconnected population in developed countries is merely 10–15%, which is significantly lower than in the developing countries where 60–70% population is still unconnected [1].

The unconnected population in developing countries is divided across gender, education levels and place of residence. The major barriers that give rise to these divides in the developing countries are (i) lack of digital awareness, (ii) unavailability of content in local languages, (iii) unaffordability and (iv) lack of internet infrastructure. Digital awareness refers to the knowledge about the internet and the benefits that it can bring about. Even if people are aware of the internet and its benefits, the unavailability of content and services in local languages discourages them from adopting the internet. Assuming the availability of relevant content and awareness of internet usage, the other two barriers of affordability and lack of internet infrastructure can be addressed with new innovations in technology. This chapter does not deal with providing solutions for content generation and awareness creation, but discusses possible technology solutions for connecting the unconnected in an affordable manner.

The global connectivity map cannot be complete without taking the Indian scenario into consideration. The current population of India is 1.34 billion of

**TV White Space Communications and Networks. http://dx.doi.org/10.1016/B978-0-08-100611-5.00007-2**

which only 291 million people are broadband subscribers [2]. Internet usage statistics suggest that approximately 317 million people access the internet at least once a month. The urban rural divide in India is even larger; only 34% of the users belong to rural areas even though 70% of the total population of India lives in these areas [3]. The Government of India, under its Digital India programme, have taken several initiatives to build an information highway connecting every part of the country. One such initiative, called BharatNet, aims to provide Points of Presence (PoPs) with optical connectivity to local self-government offices at the village level, which are called Gram Panchayats (GPs) in India. There are 250,000 GPs in India serving about 640,000 villages. Thus, each GP serves around 2–3 villages on average. These villages are generally located within a few kilometres of the GP office and will remain unconnected if fibre connectivity reaches only to the GP office. In most of the developing countries, the fibre connectivity only reaches designated points in the vicinity of the rural areas, leaving a large part still unserved. Taking optical fibre to each and every village is very time consuming and expensive. Thus, there is a need to look for alternative wireless solutions in order to efficiently connect these villages. In this chapter, we discuss wireless solutions based on Television (TV) White Space as a potential option to provide connectivity in the unserved rural areas.

In addition to providing an efficient technology solution, it is important to ensure that there is a sustainable return on investment for the service provider and an affordable subscription price for the end user. This drives the need for a sustainable economic model. We propose a model which ensures active involvement of GPs who will eventually promote and sustain the broadband. This model is referred to as the 4P model, the Private-Public-Panchayat Partnership model, which is a self-sustainable model.

The organisation of the chapter is as follows. Section 5.2 introduces the barriers in connecting the rural population. Section 5.3 discusses the TV white space as a potential option to connect the rural areas. The architecture of wireless broadband network for rural areas is described in Section 5.4. In Section 5.5, we discuss the policy aspects for the usage of TV White Space. Section 5.6 presents an economic model which can be employed in the Indian context, thus making it affordable and sustainable.

## 5.2 BARRIERS IN CONNECTING RURAL INDIA

A World Bank study suggests that a 10% increase in broadband penetration would accelerate economic growth by 1.38% [4]. Given that 70% of India's population resides in rural areas, enabling connectivity has a potential of significantly enhancing the economic growth. However, rural broadband needs to address many challenging barriers, some of which are enumerated below:

1. **Lack of fibre infrastructure**: Fibre cable is the most viable medium to support increased data capacity. China and the United States have installed around 1 billion kilometres and 600 million kilometres of fibre, respectively. In contrast, the fibre connectivity in India is very limited; only 60–70 million kilometres of fibre has been deployed to date. Currently, the rate of fibre deployment is around 15 million kilometres of fibre every year [5]. The current demand indicates the need for at least 50 million kilometres of fibre to be deployed every year. Some of the obstacles that have slowed down the process of fibre deployment are difficulties like obtaining Right of Way, uneven terrain and the cost of deployment among others.
2. **Low average revenue per user**: One of the major obstacles in enabling rural broadband connectivity is the absence of mobile network operators in these areas. The operators are not interested in serving these areas because it is not economically viable owing to a low to medium subscriber base. The Capital Expenditure (CAPEX) is significant which includes the cost of Base Transceiver System (BTS), fibre deployment and passive infrastructure. A considerable part of the Operational expenditure (OPEX) includes fuel and energy cost. Since the rural areas have low to medium subscriber density and the Average Revenue Per User (ARPU) is also low, the operators are unable to get returns on their investments, which discourages them from investing in these areas.
3. **Difficult terrain**: The difficult terrain poses a major problem in India where almost 80% of the mainland is hilly [6]. Hence, providing fibre connectivity is an arduous task in these areas. Alternate solutions like setting up a Point-to-Point wireless link (including unlicensed band radios) have been explored. However, high foliage loss and the requirement of expensive tall towers limit the feasibility of a wireless link.
4. **Scarcity of electricity**: To provide a reliable broadband access, the most important aspect is availability of electricity. Although there is electricity available, the availability in rural areas is unreliable and intermittent; there are power outages of 8–12 hours per day on average [7]. Since internet connectivity requires a continuous power supply, employing back-up power sources such as battery or diesel generators are necessary. This not only increases the energy cost, but also leads to environmental pollution. Renewable energy options such as solar energy have been explored, but have been found to be uneconomical when the power requirement of a Base Station site is high.

## 5.3 TV WHITE SPACE: A POTENTIAL SOLUTION

With the advancement of technology, multiple technology solutions, both wired and wireless, are available to provide broadband connectivity. As discussed in the previous section, providing wired connectivity in sparsely populated rural areas is an expensive solution. Wireless technologies such as IEEE 802.11 (popularly known as Wi-Fi) [9], IEEE 802.11ad (known as WiGig) [10], etc., can be suitable options. However, due to difficult terrain in India and the requirement of tall towers, these solutions will only be able to connect limited areas. In order to provide ubiquitous connectivity in such areas, a shift is required from traditional technology to a more affordable, efficient and robust technology. Wireless solutions based on TV White Space precisely suit such a scenario. In India, there is a significant amount of available spectrum in this band. In addition, it also has several unique advantages such as non-line-of-sight propagation characteristics and ease of deployment. In this section, we describe in detail the availability of TV White Space spectrum in India and the TV White Space standards suitable for rural areas.

### 5.3.1 TV White Space Availability and Usage in India

To quantify the amount of TV White Space available and the capacity that can be achieved using the available spectrum, several studies have been carried out in various countries. In the United States, only 5 channels per person are available on average in the UHF band (470–692 MHz) [13]. In Europe, around 18% of the UHF band (474–786 MHz) is available per person [14]. Similar studies have also been carried out in the United Kingdom and Japan [11,12]. These studies suggest that the availability of white space is limited and that there are only a handful of channels available per person or at any location.

The scenario is much more surprising in developing countries where there is very poor spectrum utilisation of the TV band. Unlike in the United States and the United Kingdom, there is only one terrestrial TV broadcaster in India: *Doordarshan*. Doordarshan transmits only two channels at any location in the country, each channel occupying a bandwidth of 7 MHz in VHF band or 8 MHz in UHF band. A detailed quantitative assessment of TV White Space in India reveals that 12 out of 15 channels of UHF Band IV (470–585 MHz) are available at any location [8]. Digital switchover of terrestrial TV is in progress for most of the developing countries. After digital switchover in India, the terrestrial TV broadcasting needs to shift to the 585–698 MHz band. Hence, the 470–590 MHz band will be completely vacant [15]. This unused band may prove to be very effective in connecting the rural and remote parts of the country.

Currently, there are no regulations for the usage of TV White Spaces in India. India lies in Region 3 of the International Telecommunication Union

(ITU) terrestrial spectrum allocations in which ITU permits fixed, mobile and broadcasting services in TV UHF Band [15]. As per the National Frequency Allocation Plan (NFAP) of 2011, India Remarks 36 and 37, fixed and mobile services in the 470–585 MHz band can be permitted on a case-by-case basis. Therefore, this band can be used to provide fixed wireless broadband in rural areas. This is in contrast to Europe which lies in Region 1 where only broadcasting services are allowed in this band. In the United States, which belongs to Region 2, fixed services can be permitted in the 470–512 MHz band.

### 5.3.2 TV White Space Standards

Several TV White Space standards have been developed such as IEEE 802.22, IEEE 802.11af, IEEE 802.19.1, IEEE 802.15.4m, IEEE 1900.7 and ECMA 392 [20]. Of all these standards, the two most relevant standards in providing rural connectivity are IEEE 802.22 and IEEE 802.11af. IEEE 802.19.1 is relevant in terms of the coexistence of the IEEE 802 family operating in TV White Space.

- **IEEE 802.22:** IEEE 802.22 was the first standard based on Cognitive techniques that was developed to access the TV White Spaces [16]. The most important application of this standard is to provide wireless broadband access in rural and remote areas. The standard specifies that a wide range of 33 km can only be achieved with 4 W of Effective Isotropically Radiated Power (EIRP). The standard uses both geolocation database and sensing-based techniques to access the TV White Spaces. As the Indian scenario is almost static, the use of geolocation databases can suffice. This standard is capable of working in any regulatory regime (e.g. the United States, Japan, the United Kingdom, Canada, etc.). The operational range of this band is 54–862 MHz, and it can work with various TV channel bandwidths, i.e. 6, 7 and 8 MHz.

  IEEE 802.22 follows a Point-to-Multipoint (PMP) topology with a Base Station (BS) and its associated Customer Premise Equipments (CPEs). To protect the incumbents, it follows a strict master-slave relation where BS is a master and the CPEs are its slaves. No CPE can transmit before receiving an authorisation from the BS.
- **IEEE 802.11af:** The IEEE 802.11af standard or the White-Fi was formulated to adapt the existing IEEE 802.11 for TV band operation [17]. IEEE 802.11af systems operate on frequencies below 1 GHz and use geolocation databases to access the TV bands. This standard was designed due to the congestion in unlicensed bands, i.e. the 2.4 GHz and 5 GHz spectrum. There are two operating scenarios of IEEE 802.11af: indoor and outdoor. The indoor scenario has a range of up to 100 m similar to that of Wi-Fi. The outdoor scenario has a range of about a few kilometres and is more suitable for the rural

setting. As TV channels may have varying bandwidths of 6, 7 or 8 MHz, it is required that the bandwidth be aggregated. This standard works with a bandwidth of 5, 10, 20 and 40 MHz, and depending on the availability of the channel, this bandwidth can be adapted.

## 5.4 A WIRELESS BROADBAND NETWORK FOR RURAL INDIA

### 5.4.1 Technology Requirements in Rural Areas

The current developments in cellular technology mostly focus on the requirements of high mobility. The need to support high throughput with a very high speed of mobility is driven by applications like Ultra High Definition (UHD) video streaming for users on board fast moving vehicles such as high speed trains, etc. These applications are not important in rural areas when even primary connectivity is missing. The connectivity requirements for rural areas are quite different and are listed below:

- **Low cost solution:** The average monthly income of the highest earning member of a household in a typical rural area of India is around $75 [18]. Note that a household in India comprises an average of 5 members [19]. This means that people in rural areas may not be willing to invest $3 monthly (average cost of a monthly broadband subscription with 1 GB data limit) to subscribe to broadband services. Thus, we need to develop cost-effective solutions for providing connectivity in rural areas.
- **Limited mobility support:** In a typical semi-urban or rural Indian setting, most of the end users will either be nomadic or will travel at a speed of around 30–50 km/hr. Hence, broadband connectivity at high mobility is not a major requirement in the Indian context. It is more important to focus on providing fixed broadband services with limited or low mobility in the rural parts of the country.
- **Robust and energy efficient technology:** In general, the developed countries have a very stable electricity grid, and hence its availability is not a major concern. On the contrary, the availability of electricity in rural India is highly unreliable, as described in Section 5.3. This necessitates the development of an energy efficient and robust solution which can easily work with alternate renewable sources of energy like solar energy in a cost-effective manner.
- **Support for large coverage area:** In general, rural areas are very sparsely populated. To provide ubiquitous connectivity in these areas, a huge investment in terms of internet infrastructure is needed. If Base Stations with large coverage areas are deployed, the investment can be significantly amortised, thus making the solution more affordable.

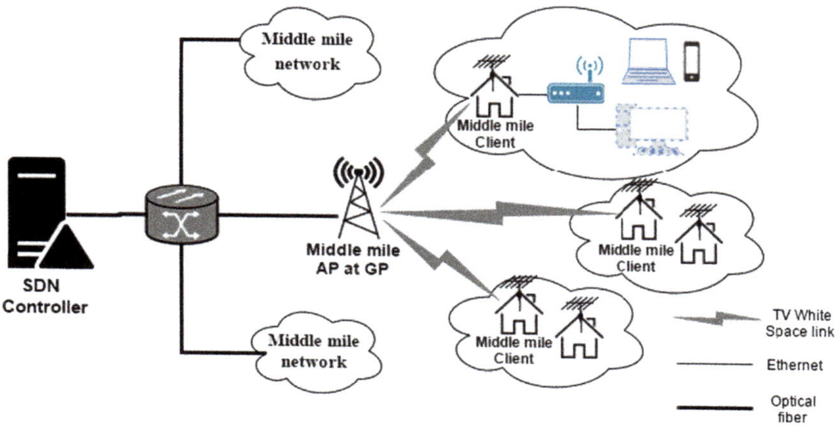


FIGURE 5.1 Rural broadband network architecture.

### 5.4.2 Rural Broadband Network Architecture

We now discuss a possible network architecture which can exploit the TV White Spaces to address the above requirements. As described in Section 5.1, the Government of India is deploying optical fibre to connect 250,000 GPs under the BharatNet project. There are 2–3 villages under each GP which are generally situated within a few kilometres of the GP office. Even after the completion of the BharatNet project, these villages will still remain unconnected as the optical fibre will terminate at the GP office. To extend the optical fibre to each and every village will be an expensive and infeasible solution. Low cost broadband access to the villages can be provided by deploying Wi-Fi Access Points (APs) as the access nodes. We need to address the issue of connecting these Wi-Fi APs to the optical PoP at the GP via TV White Space, and this network is hereafter referred to as the *middle mile network*.

Depending on the location of a GP and the villages under that GP, the middle mile network can form different topologies, i.e. (i) point-to-point, (ii) point-to-multipoint and (iii) multihop mesh networks. In Fig. 5.1, the middle mile AP locally connects to the optical PoP at the GP and the middle mile clients locally connect to the Wi-Fi APs in the villages. The end users in the villages access the internet through Wi-Fi APs. The middle mile AP is connected to multiple middle mile clients in the villages through TV White Space thus forming a point-to-multipoint topology. Various standards like IEEE 802.22 and IEEE 802.11af suit the requirements of rural areas, as discussed in Section 5.3.2, and technologies based on these standards can be effectively used in the middle mile. A middle mile Long Term Evolution – Advanced (LTE-A) network operating in TV White Space can be explored to improve spectral efficiency over the above-mentioned technologies in this band. In [21], the feasibility of using LTE-A in middle mile has been discussed. The authors in [21], propose a proportionally fair radio resource allocation over the middle mile network which uses the Coordinated Multipoint (CoMP) approach offered by LTE-A to satisfy the broadband requirement of rural areas.

To cover a large area, the middle mile topology can take a complex form including a mesh topology. A new paradigm, Software Defined Networking (SDN), can address the above problem effectively. In SDN, the control plane and data plane are separate. The control logic is completely implemented in the central controller and the network devices are simple packet forwarding devices. Therefore, it becomes much easier to manage a large-scale network such as the proposed rural broadband network.

There are several benefits attached to the proposed network architecture which are enumerated below:

1. **Reduction in cost of infrastructure:** The TV White Space has good long-distance propagation characteristics thereby providing good coverage. The coverage area of this band can be large (in kms) even with transmit power as low as 1 W. Due to the reduced power consumption, the need for diesel generators, cooling equipment, etc., can be eliminated, leading to a cost-effective solution. Another major advantage of this band is its non-line-of-sight connectivity, which will minimise the need for tall towers required for Fresnel zone clearance. In the proposed network architecture, the end users access the internet via Wi-Fi APs which are not only affordable, but also provide high speed broadband access.
2. **Ease of deployment:** When deploying Wi-Fi (5.8 GHz) APs to form a point-to-point link, antenna alignment poses a great challenge. Since the line-of-sight connectivity is not a crucial requirement with TV White Space, the installation of equipment becomes easy.
3. **Suitable for hilly terrain:** Difficult terrain in India with huge amounts of vegetation over the majority of its land poses a challenge to providing good connectivity. TV White Space can address this challenge with its non-line-of-sight propagation characteristics. Also, there is less foliage loss in this band than in higher frequencies [23].
4. **Operation using renewable energy sources:** India is a sunny region with around 250–300 clear and sunny days in a year [24]. The energy incident over India's land area is approximately 5000 trillion kWh per year, and most parts receiving 4–7 kWh per sq m per day [25]. As the power consumption of a typical TV White Space AP is less, solar power can be effectively used to power up the devices.

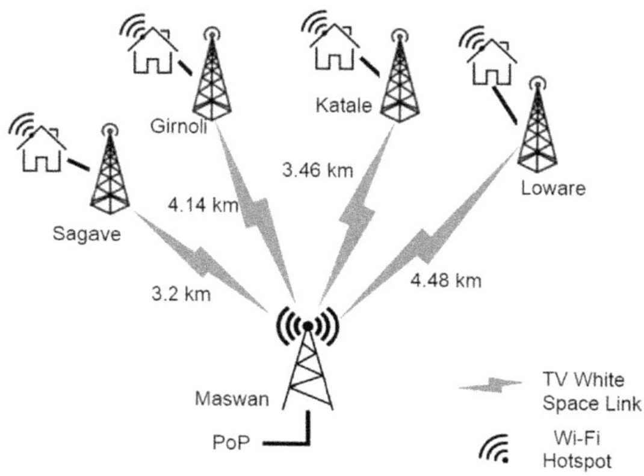


**FIGURE 5.2** An example cluster in the Palghar testbed of 25 villages.

### 5.4.3 TV White Space Testbed in India

Based on the proposed rural broadband network architecture, a TV White Space testbed has been set up in the rural areas of Palghar district, Maharashtra, India [26], which is about 100 km away from Mumbai. It includes seven villages spanning an area of about 25 sq km. The successful testbed trial in Palghar emphasises the feasibility of using TV White Space in the rural areas. Internet connectivity availed in three villages of Palghar district has provided an insight into how internet can greatly enhance the life of the villagers.

In the next stage of this testbed, it will be scaled from 7 villages to 25 villages in Palghar district. There are 5 clusters in the scaled testbed, each cluster consisting of 4–5 villages. One of the 5 proposed clusters is illustrated in Fig. 5.2. The PoP is located at Maswan village and will be connected to TV White Space AP. The TV White Space AP connects to the TV White Space clients in the 4 villages in a point-to-multipoint manner. These TV White Space clients are locally connected to the Wi-Fi APs. Owing to the long distance propagation characteristics of the TV White Space band, we will be able to connect villages which are around 4–5 km away from the PoP. This would not have been possible if unlicensed band radios had been used.

## 5.5 REGULATORY REGIMES TO ACCESS TV WHITE SPACE

It is important to note that the regulatory regimes for the use of TV White Space have not been decided in India. The behaviour of primary users in the TV White Space band is static in India. Any harmful interference to primary users can be easily avoided with the use of a geolocation database. However, the major challenge in the Indian scenario is to design a regulatory framework under which multiple operators having deployed their rural broadband network can share the spectrum efficiently to serve the rural areas.

The existing regulatory regimes can be broadly classified as *individual authorisation* schemes and *general authorisation* schemes. In individual authorisation schemes, the right of access to the spectrum is granted to an operator on a licensed basis. The individual authorisation schemes can be distinguished as *dedicated, co-primary* and *Licensed Shared Access (LSA)*. For instance, mobile operators typically have a dedicated access to the spectrum. Through spectrum auctions, they pay a very high fee and can use the spectrum for a fixed duration. Co-primary and LSA regimes have been recently introduced as complementary methods to dedicated spectrum access. These methods not only increase the spectrum utility, but operators can also ensure predictable Quality of Service (QoS) to the users. Under general authorisation schemes, users are allowed to share the spectrum opportunistically but with equal priority. The cost of spectrum is either very low or zero. For example, the Industrial, Scientific and Medical (ISM) bands are license exempt. Multiple technologies have been developed that work in these bands, e.g. Wi-Fi [9], Bluetooth [22], etc.

Given the Indian scenario where 100 MHz of TV White Space is available at any location [8], the main concern is to design a framework under which operators who are providing broadband service in rural areas can guarantee QoS to its users. The regulatory regimes under which operators can share the spectrum along with guaranteeing QoS to its users are co-primary and LSA schemes. The relevance of these schemes in the Indian scenario is discussed next:

- **Licensed/authorised shared access**: In Licensed Shared Access (LSA), the sharing takes place between primary users (e.g. TV broadcasters) and limited secondary users (e.g. mobile network operators, MNOs). The former is referred to as the incumbent and the latter is referred to as the LSA licensees. The LSA licensees are responsible for protecting the incumbent users. The spectrum sharing among LSA licensees takes place according to the rules set by the regulator. As the LSA licences are given in a limited number and all the licensees are known, it provides a higher level of guarantee of return on investment. Authorised Shared Access (ASA) is a specific case of LSA where the license holder is an MNO. ASA has been developed for the International Mobile Telecommunication (IMT) band (3.8 GHz in the United States), whereas LSA covers a much larger spectrum and is defined for heterogeneous users [27].
- **Co-primary shared access**: The co-primary shared access scheme allows sharing of the spectrum among the operators with the same regulatory status (e.g. Mobile Operators). The sharing can take place via mutual agreement among operators or enforced via requirements set by the National Regulatory

Authority (NRA). Note that in this model the operators would have equal access rights without priorities being set by regulation. As the TV White Space band is highly unused in India, a part of this band under co-primary shared access scheme can be given to operators to provide broadband services in these areas.

Co-primary sharing can be done by a limited spectrum pooling [27] method in which NRA allocates a band to multiple license holders and refrains from partitioning the band. Although it is difficult to ensure instantaneous QoS guarantees, long-term minimum QoS guarantee can be ensured via mutual agreement between operators. The mutual agreements and the number of participating licensees are known a priori. This gives some predictability to the sharing licensees, and they can get a return on their investment. Generally, the limited spectrum pooling method is discussed in the context of MNOs; however, this concept can be extended in rural areas to share the spectrum among multiple operators in the middle mile network. This band can be lightly licensed under co-primary shared access which can further aid in developing an affordable broadband solution. Therefore, it is a much more preferable scheme in the Indian rural setting.

## 5.6 ECONOMIC MODEL

In addition to providing an efficient technology solution that meets the needs of the rural areas, it is important to develop an economic model that will sustain the network and its services at the village level. This can be achieved through a sustainable economic model which is viable for both the end user and the investor. Such a model is the basis of financially self-dependent projects and its smooth running. The basic premise of an economic model is the demand and supply dynamics. Unlike the urban areas where the demand for internet connectivity and the supply by means of various technologies work in tandem, the rural areas suffer from a severe deficit. In rural areas, the demand for internet connectivity is either non-existent or it is underserved and unmet through a minimal or non-existent supply. The various reasons for this include low user rates, inefficient infrastructure, poor coverage and inadequate communication services. Communication networks, unlike road and water networks, are built on demand. With less demand in rural areas, the building of networks is slack and slow. Rural demand has also been associated with 'free' and 'subsidised' supply. It is for this reason that enabling connectivity in rural areas has not been aimed towards permanency. While the private operators are unsure of their customer base, the government operators are unsure of how to maintain and manage the networks.

A sustainable model should take into consideration the expected growth in demand in the near future. This can be achieved by (a) using economic or

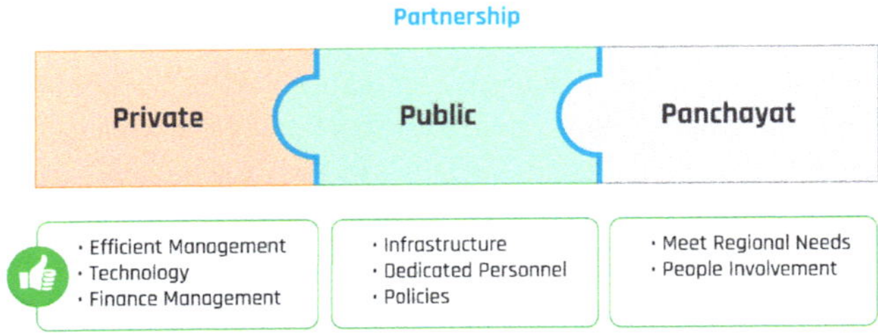


**FIGURE 5.3** 4P economic model.

cost-saving technology options so that the investment cost is low, (b) having a partnership with the government so that some part of the operating expenditure can be borne and (c) partnering with the GPs so that the local people can be involved and trained to make this model self-sustainable.

There are several types of economic models: (i) private only, (ii) public only and (iii) private-public partnership models. There have been several types of public-private partnership (PPP) models, such as the Build Operate Transfer (BOT) model, the Build Transfer Operate (BTO) model and the Joint Venture model [28]. These models have their own advantages and disadvantages. Some of the shortcomings are planning and maintenance delays, inadequate monitoring, funding gaps and improper risk management, which means that these models are unable to sustain themselves in the field for a long duration. A feature that these models lacked was the involvement of the local people for whom the network was set up and therefore did not cater to the regional needs. For example, in rural areas which suffer from maternal, child and infant deaths, the services should be oriented towards better health care facilities. Hence, we propose a 4P model: Private-Public-Panchayat Partnership model in which the Panchayat holds the responsibility of maintaining the network at the village level. In this scenario, the 'Private' could be, for instance, a private broadband service provider; the 'Public' could be, for instance, a government-owned organisation which sets up the infrastructure; 'Panchayat' is the local name for a village-level organisation which maintains and runs the network. The advantage of this model is that it is owned by the villagers themselves, which means that they will be in a better position to decide and prioritise the services that cater to their regional needs. This increases their responsibility towards maintaining and running the services in a sustainable manner. The model is depicted in Fig. 5.3. One of the key features of the model is the creation of local Village Level Entrepreneurs (VLEs) who act as a bridge between the village and the government. They are responsible for making the people digitally aware and are responsible for running these activities on a daily basis. These activities thus become a source of revenue for the VLEs, and hence are expected to be sustained for a longer duration.

## 5.7 CONCLUSION

It is an important task to enable connectivity in the unserved and underserved rural areas of India. Most of the developing countries in the world also face similar problems of connecting the unconnected. In this chapter, we have proposed a wireless broadband network based on TV White Spaces along with a sustainable economic model which can provide connectivity and can also enable successful operation of the network at the village level. However, there are a number of open problems that need to be addressed such as spectrum sharing, policy and regulation of TV White Space and universal replication of partnership business models on the ground. The proposed vision of employing TV White Space in the middle mile has the potential to radically change the current connectivity scenario and bring about a connectivity revolution.

## REFERENCES

[1] ICT Facts and Figures 2016, https://www.itu.int/en/ITU-D/Statistics/Documents/facts/ICTFactsFigures2016.pdf.

[2] http://www.trai.gov.in/sites/default/files/Press_Release_No.90_09_sep_2016_0.pdf.

[3] http://www.medianama.com/2015/11/223-317m-internet-users-india/.

[4] http://pubdocs.worldbank.org/en/391452529895999/WDR16-BP-Exploring-the-Relationship-between-Broadband-and-Economic-Growth-Minges.pdf.

[5] http://economictimes.indiatimes.com/industry/telecom/slow-fibre-rollout-hurting-telecom-health-experts-blame-indias-low-investment-in-optic-fibre/articleshow/49498198.cms.

[6] https://law.resource.org/pub/in/bis/irc/irc.gov.in.073.1990.pdf.

[7] Progress report of village electrification, https://data.gov.in/catalog/progress-report-village-electrification.

[8] G. Naik, S. Singhal, A. Kumar, A. Karandikar, Quantitative assessment of TV white space in India, in: National Conference on Communications (NCC), 2014, pp. 1–6.

[9] https://en.wikipedia.org/wiki/IEEE_802.11.

[10] IEEE Standard for Information Technology – Telecommunications and information exchange between systems—Local and metropolitan area networks—Specific requirements – Part 11: Wireless LAN Medium Access Control (MAC) and Physical Layer (PHY) Specifications Amendment 3: Enhancements for Very High Throughput in the 60 GHz Band, IEEE SA, December 24, 2012.

[11] M. Nekovee, Quantifying the availability of TV white spaces for cognitive radio operation in the UK, in: Proc. of IEEE International Conference on Communications, Jun. 2009, pp. 1–5, Workshops.

[12] T. Shimomura, T. Oyama, H. Seki, Analysis of TV white space availability in Japan, in: Proc. of IEEE Vehicular Technology Conference, Sep. 2012.

[13] S. Mishra, A. Sahai, How Much White Space Is There?, Tech. Report UCB/EECS-2009-3, EECS Department, UC Berkeley, January 2009, http://www.eecs.berkeley.edu/Pubs/TechRpts/2009/EECS-2009-3.html.

[14] J. van de Beek, J. Riihijarvi, A. Achtzehn, P. Mahonen, TV white space in Europe, IEEE Transactions on Mobile Computing 11 (2) (Feb. 2012) 178–188.

[15] National Frequency Allocation Plan 2011, WPC, Department of Telecom., Govt. of India, http://www.dot.gov.in/as/Draft.

[16] C.R. Stevenson, G. Chouinard, Z. Lei, W. Hu, S.J. Shellhammer, W. Caldwell, IEEE 802.22: the first cognitive radio wireless regional area network standard, IEEE Communications Magazine 47 (1) (January 2009) 130–138.

[17] A.B. Flores, R.E. Guerra, E.W. Knightly, P. Ecclesine, S. Pandey, IEEE 802.11af: a standard for TV white space spectrum sharing, IEEE Communications Magazine 51 (10) (October 2013) 92–100.

[18] Socio-economic caste census data, www.secc.gov.in.

[19] http://www.censusindia.gov.in/pca/Searchdata.aspx.

[20] C.S. Sum, et al., Cognitive communication in TV white spaces: an overview of regulations, standards, and technology, IEEE Communications Magazine 51 (7) (July 2013) 138–145.

[21] N.C. Prasad, S. Deb, A. Karandikar, Feasibility study of LTE middle-mile networks in TV white spaces for rural India, in: Proc. of IEEE International Symposium on Personal, Indoor, and Mobile Radio Communications (PIMRC), Valencia, 2016, pp. 1–6.

[22] IEEE Standard for Telecommunications and Information Exchange Between Systems – LAN/MAN – Specific Requirements – Part 15: Wireless Medium Access Control (MAC) and Physical Layer (PHY) Specifications for Wireless Personal Area Networks (WPANs), in IEEE Std 802.15.1-2002, pp. 1–473, 14 June 2002.

[23] K.L. Chee, S.A. Torrico, T. Kurner, Foliage attenuation over mixed terrains in rural areas for broadband wireless access at 3.5 GHz, IEEE Transactions on Antennas and Propagation 59 (7) (Jul. 2011) 2698–2706.

[24] P.K. Gupta, Renewable energy sources – a longway to go in India, Renewable Energy 16 (1999) 1216–1219.

[25] JNNSM, Mission document, Ministry of new and renewable energy government of India, Jawaharlal Nehru National Solar Mission, 2009, http://mnre.gov.in/pdf/mission-document-JNNSM.pdf.

[26] A. Kumar, A. Karandikar, G. Naik, M. Khaturia, S. Saha, M. Arora, J. Singh, Toward enabling broadband for a billion plus population with TV white spaces, IEEE Communications Magazine 54 (7) (July 2016) 28–34.

[27] http://cordis.europa.eu/docs/projects/cnect/9/317669/080/deliverables/001-METISD51v1pdf.pdf.

[28] http://www.colorado.edu/itp/sites/default/files/attached-files/70197-130943_-_gautam_kadiri_-_apr_25_2016_316_am_-_capstone_team10_spring16_final-edited.pdf.

Chapter 6

# TV White Space Spectrum Administration

**George Mastorakis*, Jordi Mongay Batalla†, Constandinos X. Mavromoustakis‡, Rossitza Goleva§, Ciprian Dobre¶ and Nuno M. Garcia‖**

**Technological Educational Institute of Crete, Agios Nikolaos, Crete, Greece, †National Institute of Telecommunications, Warsaw, Poland, ‡University of Nicosia, Nicosia, Cyprus, §Faculty of Telecommunications, Technical University of Sofia, Sofia, Bulgaria, ¶Faculty of Automatic Control and Computers, University Politehnica of Bucharest, Bucharest, Romania, ‖University of Beira Interior, Covilhã, Portugal*

## 6.1 INTRODUCTION

The pervasive delivery of multiple multimedia services and applications via emerging mobile networking systems increases the requirements for augmented access to radio spectrum resources, generating new challenges to efficiently manage the wireless networks resources and guarantee the maximum possible Quality of Service (QoS) and Quality of Experience (QoE). Several studies on radio spectrum exploitation indicate that large parts of frequency bands are not properly utilised [1], while significant parts of them are currently accessible by future wireless systems if the location and duration aspects are appropriately considered. TV White Spaces (TVWS) can be a part of such un-exploited frequency bands, including gaps in the VHF/UHF bands which were left free after the deployment of digital terrestrial television services, as well as the interleaved radio frequencies that can be accessible by future wireless systems thanks to channel planning schemes [2]. TVWS comprise great portions of frequency bands particularly at specific locations, permitting the proper deployment of energy-efficient wireless systems and the low cost provision of multiple mobile services [3]. In this context, TVWS can be efficiently exploited by emerging mobile networking systems and sophisticated wireless networks infrastructures. In order to address the request for the increasing demand of more available frequency bands, several emerging mobile systems can be exploited, like the LTE-based architectures [4] and the future 5G mobile networks. Such wireless systems enable the provision of heterogeneous multimedia services, supporting

**TV White Space Communications and Networks. http://dx.doi.org/10.1016/B978-0-08-100611-5.00008-4**

increased radio spectrum efficiency, new resource management methods and multiple transmission techniques. Furthermore, such wireless systems are flexible in using different frequency bands when both location and duration aspects are taken into account [5], co-existing with multiple current networking infrastructures.

The deployment of LTE and 5G mobile systems by exploiting TVWS will permit operators to cover a greater number of geographical locations with a small number of local base stations. Therefore, such deployment will minimise the operators' costs regarding the initial infrastructure investment, allowing for the provision of services under lower cost schemes, especially for the mobile users located in dispersed geographical areas. In addition, TVWS can be properly used to alleviate the problem of peak data traffic due to the increased numbers of requests for services from the mobile users located in urban areas. A number of different schemes to share frequency bands under an opportunistic scheme of small or intermediate periods can be adopted to support relief of congested mobile systems experiencing data peak loads. The deployment of such mobile systems needs an emerging resources management framework, appropriate for network infrastructures, where QoS has to be assured. Such a real-time framework is the RTSSM (i.e. Real-Time Secondary Spectrum Market) scheme [6,7], where spectrum trading mechanisms are adopted, in order to allow sophisticated systems to exploit spectrum resources, if the QoS is acceptable. QoS assurance is needed especially for the delay-sensitive applications and services. Such applications are multimedia-based services provided via mobile computing systems and platforms. Trading of spectrum resources exploitation takes place via resource management networking units, like a spectrum broker, implementing radio resource management (RRM) schemes [6,8], with the aim to effectively assign the existing frequency bands to secondary wireless systems [9,10]. Such wireless systems request access to the available frequency bands during a specific time period, according to need, while the payment is based on several frequency exploitation schemes, according to the radio spectrum access characteristics and to the different levels of priority among the mobile systems [11,12].

Important enablers to adopt 4G and 5G mobile systems operating under the TVWS are the Cognitive Radio (CR) systems [7,13]. CR systems allow for TVWS access by secondary wireless networks, and protect the primary networks from interference problems. In this direction, centralised networking systems [14] are suitable for 4G and 5G mobile network deployment, in comparison to distributed infrastructures, because they supply the highest possible QoS. Moreover, the adoption of a network entity like a spectrum broker allows the proper management of the available wireless resources, through the collection of data regarding the usage of the radio spectrum by primary systems and data regarding the transmission demands by the secondary wireless systems. According to such data, an optimum solution is achieved by the spectrum broker to increase the utilisation of wireless resources and the efficient management of the TVWS.

In addition, the evaluation of the wireless devices lifetime is an important issue to provide continuous multimedia services in emerging mobile systems. Such devices operate by exploiting high levels of energy consumption, while their recharging is sometimes unfeasible. They also change their location during the operation time period without guaranteeing high levels of QoS throughout the process of data exchange. This results in reduced network performance during the service delivery procedure. Therefore, energy conservation is a vital issue in future cognitive radio network architectures where network partition can regularly happen, due to the fact that the mobile nodes change their location very often. This partitioning challenge sometimes causes a portion of information to be unreachable to several cognitive radio devices. Consequently, the energy that is consumed by such mobile terminals and the resource management schemes in association with the power management policies makes it very challenging for the mobile network operators and for the radio access technology design developers to satisfy the requests for maximum QoS delivery over the entire wireless system infrastructure. As energy saving is a vital trade-off to efficiently deploy emerging mobile systems, the associated energy management frameworks must be responsive in order for the energy consumption levels of mobile devices to be adjusted, based on several parameters like data traffic and network capacity [15,16]. Generally, an energy-aware framework must consider the bounded end-to-end delays during the data exchange process. Connectivity assurance must also be provided through performance evaluation measurements regarding mobile device lifetimes, narrowly associated with the data transmission characteristics [17], as well as the primary access strategy exploited [18]. Authors in [19] developed a framework for maximising the energy conservation in CR networking architectures. This research does not consider the diversity characteristics that are affected by the temporal capacity metrics and the parameters of the proposed system. This approach also does not take into account enhanced mechanisms to avoid as much as possible the spectrum fragmentation after the allocation process of the available TVWS. An effective scheme has to associate the progressive bandwidth-aware performance of the wireless systems nodes [20] with the admission scheme exploited to preserve the energy through the mobile peer-exchange approach and through the resource access method.

To this end, this chapter progresses beyond the current research approaches by recommending two energy-aware radio resource management algorithms to efficiently exploit the TVWS and effectively preserve the energy of the nodes in a cognitive radio wireless system, according to the Streaming Path Lifetime

Maximisation (SPLM) scheme. The SPLM scheme operates in cooperation with the activity of each node in contrast to the resource sharing requests. Several network infrastructures can be exploited for deployment in specific areas by using Multi-mode Terminals (MMTs). Such terminals can be used as the recipients for a broadcasting streaming application ($S$). In this case, the streaming transfer duration is bounded by a time range as $t_1 < S_i(t) < t_2$. In addition, such terminals can be used to simultaneously access different network infrastructures and sharing network resources. The multi-radio admission policy exploited in this chapter uses a streaming scheme where the MMTs are able to exchange data streams based on the time-associated allocated bandwidth. The management of the resources of such emerging mobile systems, the economic trades of TVWS administration and the energy-aware regulator issues are arranged through a radio spectrum broker that collects data from a geolocation database related with the usage of the available TVWS. This introduction is then followed by section 6.2, which presents the design of a radio spectrum broker that is integrated in the overall CR network architecture. This section also discusses issues regarding several optimisation methods for the development of the trading and the RRM entities, presenting the suggested TVWS management algorithms. Furthermore, this part of the chapter elaborates on the throughput response enhancement through the process of tunable bandwidth adjustment to reach into an optimum energy-efficient path. Section 6.3 presents the evaluation results for the proposed energy-aware algorithms and the last part of this chapter reviews the research methods, highlighting open issues for additional research and future experimentation.

## 6.2 ENERGY-AWARE RESOURCES MANAGEMENT THROUGH AN OPTIMUM DATA EXCHANGE SCHEME

### 6.2.1 Broker-Based Energy-Efficient CR Networking Architecture

This section elaborates on the presentation of a CR network architecture that is used to effectively exploit the available TVWS based on the RTSSM policy. This CR network architecture is shown in Fig. 6.1, including a geolocation database, a spectrum broker used to manage TVWS exploitation as well as several secondary systems based on LTE technology that request access to the available radio spectrum. The spectrum broker of this architecture consists of several entities, such as the energy controller, the TVWS occupancy repository, the RRM module, the spectrum trading entity as well as the spectrum trading and policies repository. The basic role of the energy controller is to allocate energy-aware paths according to the data exchange requests by the Secondary Nodes (SN). This is evaluated via the Streaming Path Lifetime Maximisation (SPLM) scheme that is assessed by every energy module to preserve energy

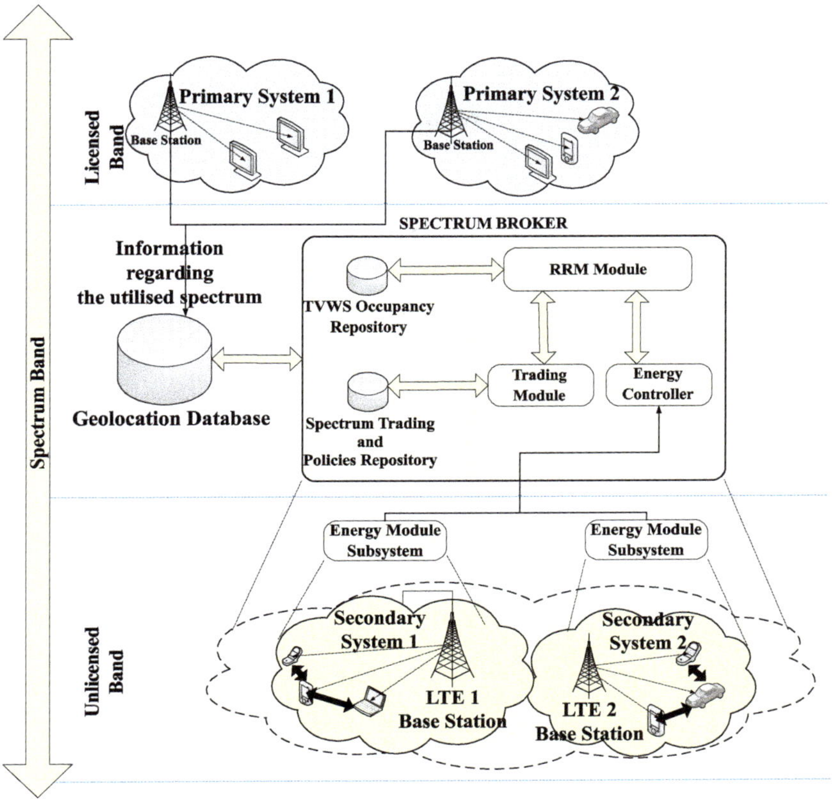


**FIGURE 6.1** Broker-based network architecture enabling TVWS management.

levels and assign the optimum energy-efficient path through a peer-to-peer resource exchange method or the RRM module. The RRM module manages the demands for wireless resources by the SN and the available radio spectrum access, assigning the TVWS according to the explicit QoS requests. This module implements RRM processes, by using records which are provided by the database. It is used to compute all the frequency slots that are available and the transmission power maximum thresholds allowed to be used by secondary systems without causing interference and spectrum fragmentation. This module is also used to optimise the RRM process and support QoS provision or guaranteed fairness to TVWS access. In addition, the spectrum trading/policies repository contains data useful for the TVWS trading procedure. This repository also includes information regarding the price per part of spectrum and creates a price portfolio. The TVWS occupancy repository gathers all requested data from the database about the availability of frequency slots in specific areas. This repository generates a spectrum resources portfolio that includes such information for the secondary systems. Finally, the trading module calculates the profit of the spectrum broker that leases the available radio spectrum to the most valuable secondary systems.

### 6.2.2 Algorithms for Efficient Exploitation of TVWS

The spectrum broker in Fig. 6.1 is responsible for trading the accessible bandwidth, between several secondary systems which are involved in the procedure of the TVWS allocation. This entity advertises the available parts of the radio spectrum for trading and all related data for the maximum levels of the allowable transmission power. For this reason, it informs all secondary systems about the spectrum and price portfolio that includes the related data for each part of the existing TVWS. After this phase, the mobile systems request their claims regarding the offered radio spectrum parts and the broker initially gathers such demands, analyses them and processes them based on several technical requirements. For every portion of the available frequency slots, the spectrum broker keeps a record with the related requests from the secondary systems. This is a request portfolio, used to assign every frequency band part to the most profitable mobile systems, considering the QoS constraints such as the priority levels.

In addition, the combination of the procedures of the RRM and trading modules enables the proposed system to effectively allocate the available frequency parts to the most valuable cognitive radio systems. This decision can be reached according to an optimisation process that tries to keep the spectrum fragmentation as low as possible in the case of the fixed-price approach. On the other hand, it tries to increase the benefit of the system in the case of the auction-based approach. For this reason, the proposed resources management entity exploits optimisation processes [21] that consider decision making approaches to reaching a solution through classical mathematical rationalisation. In this context, several methods can be adopted to support the optimal solution, like Backtracking, as well as Genetic Algorithm or Simulated Annealing. The former approach finds a global optimal solution by considering several sets of solutions, while the latter chooses it from among a reduced set of solutions, respecting the objective function.

In this framework and to elaborate on the fixed-price approach, the proposed Algorithm 6.1 is exploited to obtain an optimum solution. This is achieved by elaborating on the "$C(A')$" (i.e. objective function in Eq. (6.1)), regarding the prioritisation ($Pr(i)$) for the cognitive radio systems, the fragmentation ($Frag(i, f)$), the requested bandwidth ($BW(i, f)$) and the transmission power ($P(i, f)$). The target in the fixed-price method is to keep the radio spectrum fragmentation as low as possible to prevent slight parts of frequency bands from

**Algorithm 6.1:** Pseudo-Code.

1: **Initialise:** Secondary Systems Request, Max. Transmission Power, Coordinates $(x, y)$
2: **Initialise:** Price per MHz
3: Update TVWS repository from the geolocation database
4: Create and advertise price portfolio
5: Set request of Secondary systems $R = \{R_1, \ldots, R_i\}$, where $R_i = \{x_i, t_f\}$
6: **for** all Requests
7:     Descending sort of request "$R_i$" considering priority
8:     Update price portfolio
9: **end for**
10: Calculate the Min. fragmentation score $(Frag(i, f))$
11: Construct initial solution $S$
12: **for** $i = 1$ to subset of variable length
13:     Construct neighbour solution $S_i$
14:         **if** (solution $(S) \leq$ solution $(S_i)$)
15:             **then** $S = S_i$ // best found solution
16:         **end if**
17: **end for**
18: **find a path** $T_{SPLM_i}$ with $\min(T_R, T_{i \to j})$
19: if $T_R < T_{i \to j}$ and $B_{i \to j} \leq B_R$
20: **estimate** $\arg\max_{\delta_{ij} < d_p} \Theta(t_s) | T_i \forall \min(T_R, T_{i \to j})$
21: **allocate** $S_i$ for $t_s, \forall t < S_i(t)$
22: **return** Best Allocation Solution

existing after the allocation process. $P(i, f)$ in Eq. (6.1) denotes the maximum allowed power of "$f$" TVWS channels that can satisfy the technical requirements of secondary systems, while $BW(I, f)$ represents the bandwidth that is requested by a secondary system. In addition, the SPLM scheme is exploited to assign the optimum communication path to the mobile terminals based on the energy consumption levels. This results in an energy-efficient data delivery procedure that is adopted through the radio spectrum broker or through a peer-to-peer resource exchange approach.

$$\text{min:} \quad C(A') = \sum_{i \in V} \sum_{f \in F} x_{if[P(i,f)+BW(i,f)+Frag(i,f)+Pr(i)]} \qquad (6.1)$$

Algorithm 6.1 is exploited under a fixed-price approach to minimise the objective function and obtain an allocation solution "$X$" for a secondary system "$i$" that requests access of "$f$" available TVWS channels. For this purpose, Algorithm 6.1 takes into account a number of criteria. It first takes into account $BW(i, f)$ and $P(i, f)$ requirements from secondary systems. It then considers several criteria regarding the QoS of secondary systems. $Pr(i)$ defines the priority level that a secondary system is associated with to access the available radio spectrum. In addition, $Frag(i, f)$ calculates the percentage of spectrum fragmentation after the allocation process. This percentage has to be as small as possible. Finally, "$R_i$" in the algorithm below is the request of an "$i$" secondary system for a time period "$t$" to access "$f$" TVWS.

In addition, the proposed scheme takes into account the MMT arrangement, as well as the Radio Access Technology, to evaluate the minimum possible energy path as depicted in Algorithm 6.1. The MMTs demand delay-sensitive streams during a specific time frame. In the other case, they will be prohibited. This chapter exploits $M$ terminals of MMTs type, which demand delay-sensitive and transmission-bounded delay services. *R Radio Access Technology terminals* $(R < N)$ are set as presented in Fig. 6.1. The streaming process is bounded by delay-constraint service traffic for the MMTs in which the requested missing streaming chunks are requested by other Radio Access Technology terminals in order to be able to assemble them in the end-terminal MMT. Algorithm 6.1 shows that if the residual energy is higher for MMT devices then the requested resources are being transferred via the MMTs or otherwise through the $R$ Radio Access Technology terminals. To this end Algorithm 6.1 evaluates and estimates the remaining energy on devices of MMTs and the available bandwidth of each path selected either by using the MMTs to transfer the requested resources or by using the Radio Access Technology terminals, provided that the transfer delay is within the allowed bounded $t_s$ for the $s$ Stream to adequately reach the destination. A similar evaluation is performed in Algorithm 6.2, where the available bandwidth of each path selected is being evaluated in time in contrast to the residual energy of each MMT in the relay path.

On the other hand, the spectrum broker is able to operate based on the auction-based approach and the proposed Algorithm 6.2. In this case, the spectrum broker receives the bids from the secondary systems to initiate an auction process for the TVWS leasing. It then produces the result of the allocation procedure by considering the price for every part of the available radio spectrum. To maximise the profit for the spectrum broker entity, the optimisation approach was formulated:

$$\text{max:} \quad \sum_{k=1}^{k} \sum_{k=1}^{k_n} \left(P_i^{(b)} - P_n^{(s)}\right) x_{i,n} \tag{6.2}$$

**Algorithm 6.2:** Pseudo-Code.

**Initialise:** Secondary Systems Request, Max. Transmission Power, Coordinates $(x, y)$
**Initialise:** Price per MHz
Update TVWS repository from the geolocation database
Create and advertise price portfolio
Receive bids from secondary systems $P^{(b)} = \{P_1^{(b)}, \ldots, P_I^{(b)}\}$, where $P_i^{(b)} = \{x_i, t_f\}$
**for** all Bids
    Descending sort of bids "$P_i^{(b)}$" considering price
    Update price portfolio
**end for**
Calculation of the maximum benefit $S[i, s]$, where $(i, s) \ni \{1, 2, \ldots, m\}$
set $S_{optimal} = S[i, s]$ //Random solution for the initiation of the algorithm
**for** slot $= 1$ to $m$ //Iteration process to find the best solution
    **if** $(S[i, s]) \leq (S[i+1, s+1])$ // Check if the current solution is better than the neighbour solution or not
        **then** $S[i, s] = S[i+1, s+1]$
    **end if**
**end for**
**find a path** $T_{SPLM_i}$ with $\min(T_R, T_{i \to j})$
if $T_R < T_{i \to j}$ and $B_{i \to j} \leq B_R$
**estimate** $\arg\max_{\delta_{ij} < d_p} \Theta(t_s) | T_i \forall \min(T_r, T_{i \to j})$
**allocate** $S_i$ for $t_s$, $\forall t < S_i(t)$
**return** Best Solution

where each buyer "$i$" (i.e. mobile systems) requests buying $x_i$ parts of frequency bands by sending a price $P_i^{(b)}$ (Bid Price), while each seller "$n$" (in our case $n = 1$, the spectrum broker) sells $y_n$ parts of radio spectrum by reporting a price $P_n^{(s)}$ (Asking Price). $k$ is the total number of frequencies (i.e. the TVWS which can be traded by secondary terminals). Finally, $x_{i,n}$ is the amount of spectrum which is purchased from the spectrum broker by "$i$" secondary systems. When the auction-based algorithm is adopted, the frequency suppliers are referred to as $n$, where $n = \{1, 2, \ldots, n\}$; the available spectrum is denoted $S = \{1, 2, \ldots, s\}$ and the secondary systems are denoted $i$, where $i = \{1, 2, \ldots, i\}$. Each cognitive radio system "$i$" can obtain "$x_i$" parts of TVWS for exploitation during a

**TABLE 6.1 Notations Used in the Proposed Framework**

| Symbol | Representation |
|---|---|
| $B_{i \to j}$ | Frequency bandwidth allocated to $\text{MMT}_i$ from $i$ node to $j$ node |
| $T_i$ | Lifetime of MMT through either $R_i$ or the $\text{MMT}_{i+1}$ for $k$-hops |
| $t_s$ | Upper allowed end-to-end delay for the $s$ Stream to reach the destination |
| $S_i$ | Multimedia stream chunks for distribution to $\text{MMT}_i$ |
| $\Theta(t_s)$ | Throughput for a given nodal lifetime $T_i$ |

time period "$t_i$". This is achieved by reporting a price $P_i^{(b)} = \{x_i, t_i\}$. "$(i, s)$" in Algorithm 6.2 represents all the different solutions to match "$s$" TVWS to "$i$" spectrum buyers. Furthermore, "$P_i^{(b)}$" in Algorithm 6.2 denotes the bid of "$i$" secondary systems, requesting access to TVWS channels for a time period of "$t$", while "$m$" is the sum of TVWS slots that are available for the auction process. Notations utilised by both Algorithms are introduced in Table 6.1.

### 6.2.3 Secondary Terminal Resource Exchange, Using Delay-Constrained and Terminal-Aware Energy Model

This part of the chapter presents a scheme to explore an energy-aware path among networking nodes by exploiting a Subsystem Energy Module as depicted in Fig. 6.1. The proposed use-case approach exploits a number of SNs that are located on a specific geographical range. SNs receive a joint distribution resource exchange $S$ application where a time delivery constraint exists, which is limited in $t_1 < S_i(t) < t_2$. The multi-radio admission approach that is exploited in this scheme employs a resource exchange strategy, where the SNs are able to receive and distribute data streams based on the time-oriented allocated bandwidth. According to this configuration, the SNs can simultaneously deliver the demanded content via parallel transmission [20]. In the case of fading characteristics, the instantaneous Channel State Information (CSI) matches the conditions of each channel, occasionally updating information in the presence of response collaboration and delivery between channels. The proposed research approach exploits a model that is able to take into account this configuration by using the SNs and by using the Cognitive Radio networking architecture via the RRM procedure (i.e. Fig. 6.1). Based on this, the SNs request delay-sensitive data at no regular intervals that is required among a specific time frame, otherwise they will be rejected. According to the proposed scheme, $M$ devices of SN type are exploited demanding delay-sensitive and transmission bounded delay services; and $R$ *Cognitive Radio Access Technology terminals* ($R < N$) are set as presented in Fig. 6.1 (i.e. LTE Base station(s)). The main target is to discover an efficient transmission power assignment of cooperative SN nodes that is able increase the lifetime of the SNs. This lifetime is defined as the moment when the first node is depleted of energy [20]. If the residual energy and the consumption of the transmission power of a node $i$ in the set of $M(i \in M)$ are $E_i$ and $P_i$, respectively, then the lifetime of node $i$ is measured as:

$$T_i = \frac{E_i}{P_i} \tag{6.3}$$

In this framework, Streaming Path Lifetime Maximisation (SPLM) policy is exploited to assign a specific path among the Cognitive Radio networking architecture, in such a way to increase the lifetime of the SN by using the $R(k)$ and the cooperative $k$-hop relay path. This is depicted in Fig. 6.1 in both a peer-based approach and through the CR base-station $R(k)$. According to this approach, the spectrum broker collects data demands by the SN terminals to investige the most energy-aware path. The Energy Controller allocates an energy-efficient data delivery exchange path to the SNs based on the related subsystem energy module to which the SN belongs. The energy efficiency is assessed by each secondary subsystem energy module, exploiting the SPLM model to conserve energy. This is done by assigning the optimum energy-efficient path, either via the RRM module or by exploiting the peer-to-peer resource exchange procedure. In the case of losses throughout the data delivery process—as in [22]—a backup node can be set in the vicinity of the $SN_i$ such that $SN_{i+1}$ can claim the chunk streaming resources, based on a promiscuous caching policy [23]. In this context, the SPLM can be defined as

$$T_{SPLM_i} = \min(T_R, T_{i \to j}) \tag{6.4}$$

where $T_R = \frac{E_R}{P_R}$ and $T_i = \sum_i^j \frac{E_i}{P_i}$ are respectively the lifetime of SN through $R_i$ and through the $SN_{i+1}$ for $k$-hops, from which $SN_i$ can claim the chunk streaming sharing resources based on the promiscuous caching purposes. This evaluation happens until reaching the destination $j$.

The SPLM is effective from the perspective of uninterrupted data exchange flow through the streaming path. To this end, the power assigning process can be formulated, under Bit Error Rate (BER) constraint as:

$$n_i \forall R_i > T_R \tag{6.5a}$$

$$n_i \forall \max\left(\sum_i^j \frac{E_i}{P_i}\right) \in M \tag{6.5b}$$

where $n_i$ is the BER for exploiting the RATs and Access Points (Eq. (6.5a)) and the peer accessing nodes $SN_i$ to $SN_{i+1}$ via $k$-hop SNs. Eqs (6.5a) and (6.5b) show respectively the estimation of the minimum energy selection path for streaming through $R_i$ and through the cooperative relaying via Mobile Peer-to-Peer resource sharing. Eq. (6.5b) shows that if the streaming relay path is selected through the peer users then the lifetime of each device in the relay path should be maximised. This means that the minimum energy should be consumed if the relay path $i$ to $j$ is selected. The basic target is to perform the minimum transmission power allocation of the cooperative nodes that could maximise the lifetime of the network for an end-to-end BER below the allowed $BER_{\max}$. Therefore, the $BER_i$ should satisfy the upper limit for the BER in the end-to-end path threshold as follows:

$$\sum_{\substack{i=1 \\ SPLM_i>0}}^{N} BER_i < BER_{\max} \tag{6.5c}$$

In order to obtain a specified level of the desired Throughput $\Theta(t_s)$ for a given nodal lifetime $T_i$, where for all the $k$ streaming pieces it stands that $S_i = \{\sum_1^k S_k\}\ \forall k \in j, R$ and $S_i(t), \forall t_1 < S_i(t) < t_2$ (each stream should be adequately transferred within a specified time duration, otherwise the stream is incomplete and intermittently gradually missed), then the $\Theta(t_s)$ can be evaluated via

$$\Theta(t_s)|T_i = \max\left(\sum_{i}^{j,R}(1-n_i)\sigma_i B_{i\to j}\log_2\left(1+\frac{g_{ij}\cdot P_{ij}}{B_{i\to j}}\right)\right) \quad \forall S_i \in j, R \tag{6.6}$$

Here $B_{i\to j}$ is the bandwidth that is assigned to $\mathrm{SN}_i$ from the RAT or to the neighbouring $\mathrm{SN}_i$ peer, $\sigma_i$ is the streaming parameter as in [22], $n_i$ is the BER for exploiting the RATs and Access Points and the peer retrieving nodes $\mathrm{SN}_i$ to $\mathrm{SN}_{i+1}$ via $k$-hop SNs, $P_{ij}$ is the transmission power from and to each $\mathrm{SN}_i$ where it stands that $\forall P_i = 1 - T_{SPLM_i}$, $i \in M$ and $g_{ij}$ is the channel gain. The transmission power $P_{ij}$ can be evaluated as the total path transmission power with $P_{ij} = \sum_i^j 1 - T_{SPLM_i}$.

The delay that the transmission experiences $\delta_{ij}$ should satisfy $\delta_{ij} < d_p$, where $d_p$ is the maximum end-to-end delay:

$$d_p = \sum_{i=0}^{i-1}\delta_i + T_i \tag{6.7}$$

In this case, $\delta_i$ represents the period passed for the information to be hosted in the $i$-node. $T$ denotes the communication delay. To attain reduced energy consumption levels in the path $E_{C_i}$, the following has to be fulfilled:

$$\underset{\delta_{ij}<d_p}{\arg\max}\,\Theta(t_s)|T_i = \left\{\delta_{ij} < d_p : \Theta(t_s)|T_i = \min f(T_{SPLM_i})\right\} \tag{6.8}$$

## 6.3 PERFORMANCE EVALUATION ANALYSIS, EXPERIMENTAL RESULTS AND DISCUSSION

### 6.3.1 Experimental Test-Bed Description and Technical Specifications

A decision making process was developed by using Simulated Annealing [21] to test the efficiency of the proposed energy-aware scheme and of the effectiveness of the algorithms and the capacity of the Cognitive Radio networking architecture for optimal TVWS allocation among SNs. In this framework, several simulation tests were conducted to assess the proposed algorithms in terms of the proposed system profit and of the optimum residual energy-efficient path for the data delivery procedure. The simulation tests were based on several performance evaluation metrics to investigate the optimum values which are able to support an energy-aware resource management. The experimental test-bed comprises a TVWS occupancy repository that possesses records regarding the television bands which are able to be exploited by the LTE systems of the proposed CR networking architecture. Data inside such a repository was collected from actual radio spectrum information according to the results of the "CogEU" project [24]. This data concerns the TVWS accessibility between 626 MHz and 752 MHz in Munich, Germany. It should be mentioned here that based on the experimental tests, the fixed-price and the auction-based approaches were both selected according to a single spectrum-unit price. This experimental use-case comprises seven LTE systems with multiple radio characteristics which are concurrently contending for the obtainable TVWS within four time durations.

### 6.3.2 Simulation Results and Discussion

In order to examine the proposed framework from different performance perspectives we conducted multiple simulation experiments using discrete-time simulator. The experimental part aims to provide an objective quantitative and qualitative comparison for the proposed RRM algorithms. The conducted experiments focus on the response of the system, and more specifically on issues regarding the energy consumption of the SNs. Moreover it investigates the benefit of the spectrum broker and the overall contribution in the offered throughput of the system. The evaluations aim to show the performance comparisons of the proposed scheme compared with other approaches in contrast to the obtainable throughput, and also shows the degree of reliability when delay-sensitive transmissions are hosted in the system. The main evaluation criterion is to tune the nodes' channel bandwidth so that the energy conservation is maximised (Eq. (6.6)) during the process of delay-sensitive transfers. To evaluate the proposed framework, this work uses a probabilistic Fractional Brownian Motion

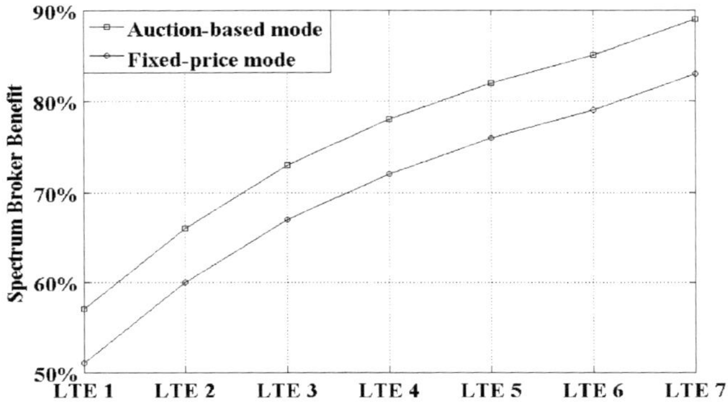


**FIGURE 6.2** Spectrum broker benefit.

(FBM) [23], where SNs are moving in any direction based on the probabilities associated with the location of the SN throughout the prior time slot and with the corresponding time. This work uses the simulated topological model, as described in [15], where the topology is based on a 'grid-based' network [16]. In this approach, every node directly contacts other nodes in the case that the node in reference is within a proximity area (i.e. $3 \times 3$ centre). The varying parameters described above were exploited to develop a networking system environment that comprises of maximum 200 $SN_i$ (terminal nodes that are also acting as relay nodes), which are situated in the measured area (<50 for each area/subsystem).

In this context, Fig. 6.2 presents the radio spectrum broker benefit for the auction-based and the fixed mode RRM algorithms. It is observed that the radio spectrum broker benefit grows with the increase in the number of the LTE secondary systems that simultaneously access the available TVWS channels. The auction-based algorithm delivers an enhanced performance regarding the spectrum broker benefit, in contrast to the fixed-price algorithm.

In addition, the offered throughput response in contrast to the SNR (Signal-to-Noise Ratio) (dBm) and the simulation duration time are depicted in Fig. 6.3 and Fig. 6.4, respectively. It is shown that the throughput response is considerably higher by using the proposed scheme to allocate the optimal energy-aware path to the SNs during the data delivery process. Fig. 6.5 depicts the lifespan of the SNs for both the assessed schemes. It is shown that the suggested method supports increased lifetime extensibility, when the joint $R(k)$ and SNs peer streaming is exploited. Moreover, Fig. 6.6 depicts the energy efficiency (kbit/J) with the delay request, where it is verified that the suggested joint scheme supports an important reduction in the delay, even when the energy efficiency is greater. This is due to the SPLM method used that allows the allocation of the

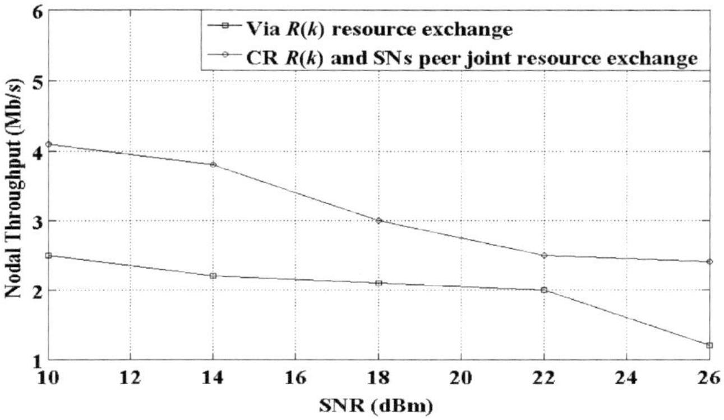


**FIGURE 6.3** Nodal Throughput with SNR (dBm).

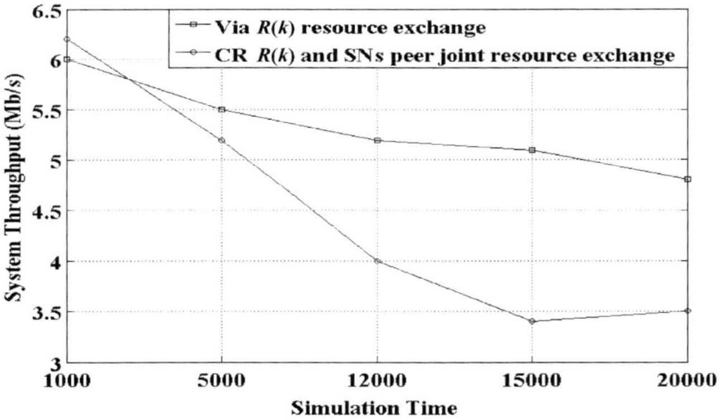


**FIGURE 6.4** System Throughput during simulation.

expected energy-aware path to SNs to fulfil the delay requests during the data delivery process.

Moreover, the total energy consumption with the number of participating SNs for three alternative schemes is depicted in Fig. 6.7. It is absolutely true that the energy consumption is considerably decreased when both the $R(k)$ and the SNs data delivery are exploited. In addition, when the P2P method is exploited the energy consumption is marginally greater. Fig. 6.8 presents the Complementary Cumulative Distribution Function (CCDF) transmission sharing reliability with the number of SNs for alternative secondary subsystem participation, depicting that when the number of SNs increases, the resource sharing reliability increases, offering higher data sharing rates.

This work uses the Link Utility Maximisation (LUM) mechanism, which is based on the link that each $SN_i$ can claim chunk streaming sharing resources

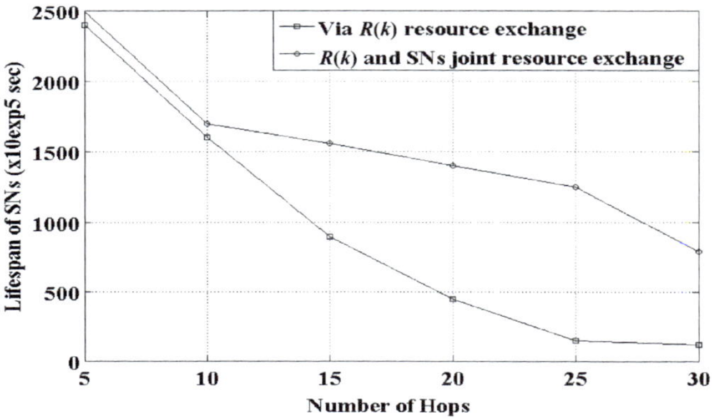


**FIGURE 6.5** Lifespan of SNs for both evaluated schemes.

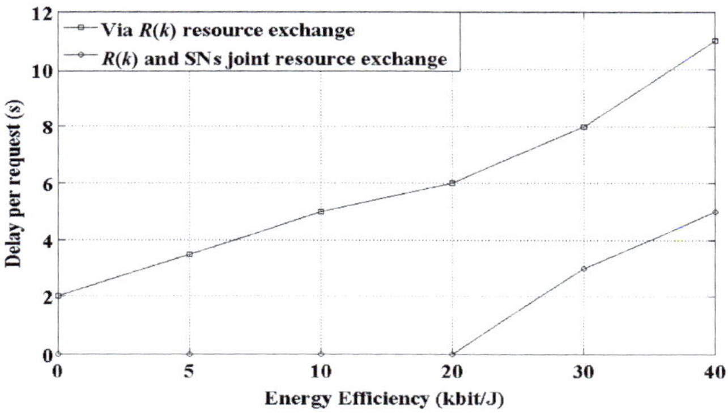


**FIGURE 6.6** Energy efficiency (kbit/J) with delay request.

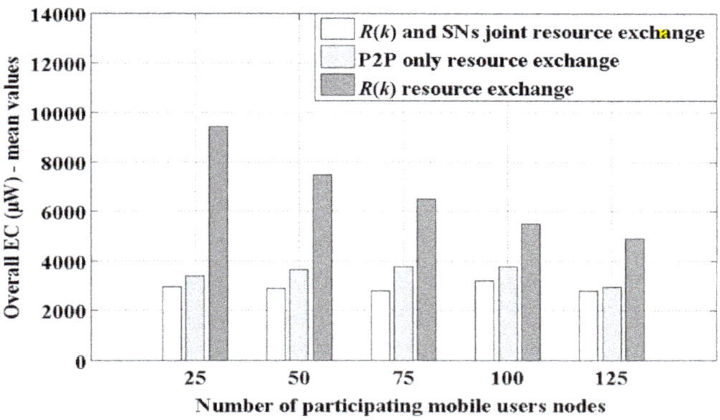


**FIGURE 6.7** Overall energy consumption for each SN for three different schemes.

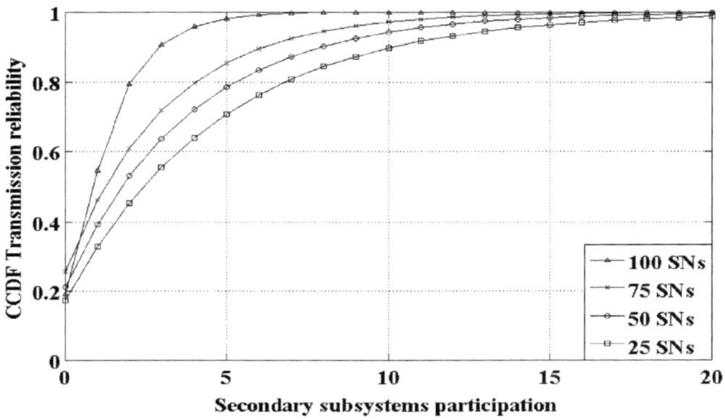


**FIGURE 6.8** CCDF transmission sharing reliability.

according to the utilisation degree of each link, and should satisfy the following:

$$\Lambda_{LUM} : \max \sum_{r \in R} U_r(\chi_r) \quad \forall \mathrm{A}_\chi \leq C,\ \chi \geq 0 \tag{6.9}$$

In this case $r$ represents the index of source node (i.e. $\mathrm{SN}_i$), $\chi_r$ denotes the transmission rate of $\mathrm{SN}_i(r)$. Also, $C$ is the vector that includes all associated capacities for the links of $\mathrm{SN}_i(r)$ and, finally, $U_r(\chi_r)$ represents the utility of node $r$ after transmitting, exploiting a rate $\chi_r$. According to Eq. (6.9), the link utility denotes the availability degree of a node as a matter of transmission rate. In addition, $\mathrm{A}_{\chi(r)}$ represents a matrix containing the associated paths (from and to $\mathrm{SN}_i(r)$) which are denoted 1 if there is such route (via $R_i$ and $\mathrm{SN}_i$) and 0 otherwise. In this respect, the SDR (i.e. Successful packet Delivery Ratio) versus the concurrent flows-requests of the Secondary System is presented in Fig. 6.9 under different LUM parameter values. As the LUM parameter is based on the link utilisation through which each $\mathrm{SN}_i$ can claim file-chunk sharing resources, it indicates the degree of utilisation, so that each $\mathrm{SN}_i$ can follow the certain link. This is performed through the LUM policy which enables a vector, which in turn reveals the associated capacities of all the links. It is important to point out that by using $R(k)$ resource exchange policy in the presence of high LUM values for low concurrent transfer measures, the SDR is kept to lower levels. On the contrary, by utilising joint $R(k)$ and SNs resource exchange SDR increases significantly.

Fig. 6.10 depicts the network size for different numbers of completed chunks and Fig. 6.11 shows the end-to-end latency with the number of mobile users (SNs) during the simulation process. Fig. 6.10 also presents the associated network size and Throughput response (vertical dashed lines) of the Secondary System with the number of completed files and network size. When network size

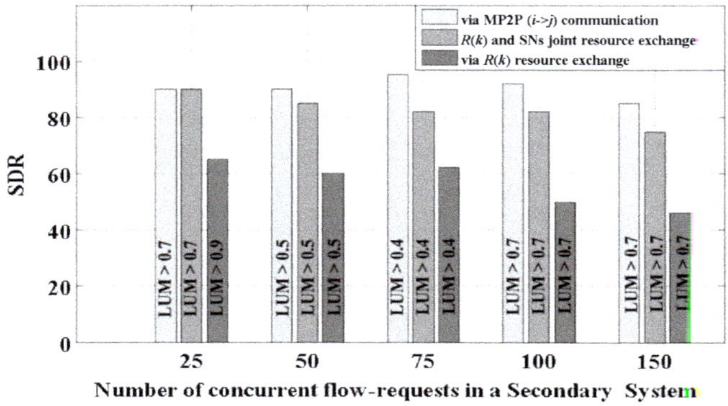


**FIGURE 6.9** Successful packet Delivery Ratio (SDR).

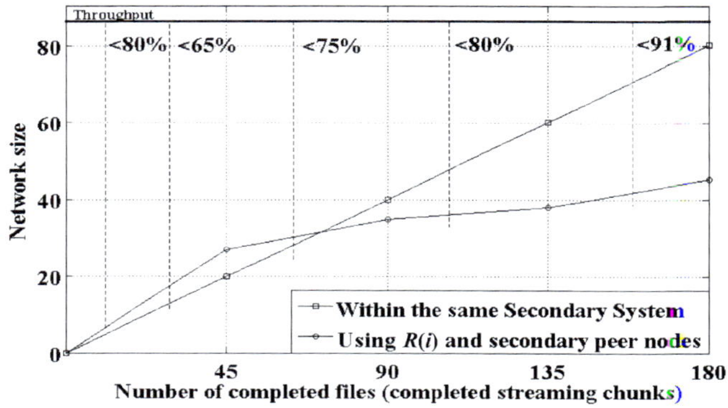


**FIGURE 6.10** Network size for different numbers of completed chunks.

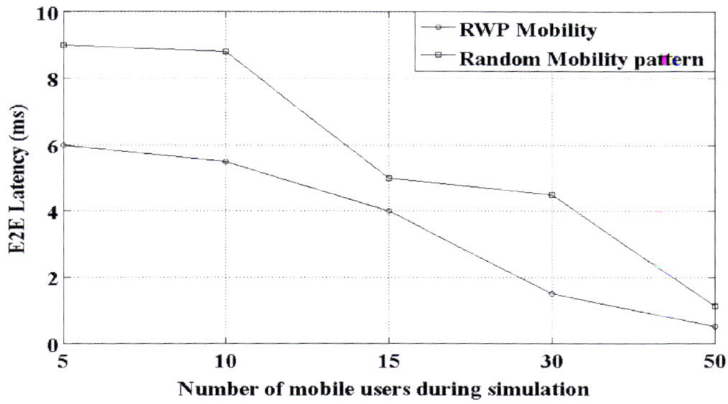


**FIGURE 6.11** End-to-end latency with number of mobile users during simulation.

increases, Throughput response increases according to the number of completed files. This occurs due to the evaluations of Eq. (6.6) that take place for estimating the desired Throughput $\Theta(t_s)$, for a given nodal lifetime $T_i$. In turn the End-to-End latency is reduced when the number of SNs in the system increases. This is due to the joint $R(k)$ and SNs resource exchange which enables alternative paths to be followed and according to the extracted results in Fig. 6.11, the latency is reduced for both the mobility models used.

## 6.4 CONCLUSION

This chapter proposes two algorithms that are able to be adopted in a Cognitive Radio networking architecture to support efficient TVWS utilisation, as well as consistent data delivery through optimum energy-efficient paths under the RTSSM scheme. The work presented in this chapter also elaborates on the development of an energy-aware radio spectrum broker by exploiting decision making methods based on Simulated Annealing. Several experimental tests were conducted to evaluate the performance of the two algorithms, where cognitive radio systems concurrently exploited the available frequency slots among multiple time durations. The recommended mechanisms were enhanced by an energy module to offer optimal energy-aware path assignment to the SNs. The experimental results confirmed the efficiency of the radio spectrum broker to optimally allocate the available radio spectrum resources under an energy-efficient approach during the resource exchange process of the SNs. Several open issues were identified after the successful completion of this research work, which will be investigated in the future, including a quantitative and qualitative comparison with other optimisation algorithms/mechanisms. Future research also includes the expansion of the proposed scheme to an energy-efficient motion-aware streaming methods, exploiting cognitive radio and peer-to-peer data resource exchange approaches.

## ACKNOWLEDGEMENT

This work was funded and supported by the ICT COST Action IC1303-AAPELE, Architectures, Algorithms and Platforms for Enhanced Living Environments.

## REFERENCES

[1] First report and order, Federal Communication Commission Std, available: http://fjallfoss.fcc.gov/edocs_public/attachmatch/FCC-10-196A1_Rcd.pdf, 2010.

[2] The Economics of Spectrum management, A Review, Australian Communication and Media Authority (ACMA), 2007.

[3] OFCOM, Digital Dividend Review: geographic interleaved awards 470–550 MHz and 630–790 MHz – Consultation on detailed award design, June 2008.

[4] 3GGPP LTE, http://www.3gpp.org/LTE.
[5] Unlicensed Operation in the TV Broadcast Bands, Final Rules, available: http://edocket.access.gpo.gov/2009/pdf/E9-3279.pdf, last accessed, Aug. 2013.
[6] A. Bourdena, E. Pallis, G. Kormentzas, C. Skianis, G. Mastorakis, Real-time TVWS trading based on a centralised CR network architecture, in: Proc. IEEE Globecom2011, Texas, Houston, USA, 2011, pp. 964–969.
[7] I.F. Akyildiz, W.Y. Lee, M.C. Vuran, S. Mohanty, A survey on spectrum management in cognitive radio networks, IEEE Communications Magazine 46 (4) (2008) 40–48.
[8] F. Cavalcanti, S. Andersson, Optimizing Wireless Communications Systems, Springer, 2009.
[9] A. Bourdena, E. Pallis, G. Kormentzas, G. Mastorakis, A prototype cognitive radio architecture for TVWS exploitation under the real time secondary spectrum market policy, Physical Communication 10 (2014) 159–168, http://dx.doi.org/10.1016/j.phycom.2013.11.003.
[10] A. Bourdena, E. Pallis, G. Kormentzas, G. Mastorakis, A centralised broker-based CR network architecture for TVWS exploitation under the RTSSM policy, in: Proc. IEEE ICC2012, Ottawa, Canada, 2012, pp. 7243–7247.
[11] A. Bourdena, E. Pallis, G. Kormentzas, H. Skianis, G. Mastorakis, QoS provisioning and policy management in a broker-based CR network architecture, in: Proc. IEEE Globecom 2012, Anaheim, California, USA, 2012.
[12] A. Bourdena, E. Pallis, G. Kormentzas, G. Mastorakis, Radio resource management algorithms for efficient QoS provisioning over cognitive radio networks, in: Proc. IEEE ICC2013, Budapest, Hungary, 2013.
[13] A. Wyglynski, M. Nekovee, T. Hou, Cognitive Radio Communications and Networks: Principles and Practice, Academic Press, 2009.
[14] E. Hossain, D. Niyato, Z. Han, Dynamic Spectrum Access and Management in Cognitive Radio Networks, first ed., Cambridge University Press, 2009.
[15] C.X. Mavromoustakis, C.D. Dimitriou, G. Mastorakis, Using real-time backward traffic difference estimation for energy conservation in wireless devices, in: Proc. of 4th International Conference on Advances in P2P Systems (AP2PS 2012), Barcelona, Spain, 2012, pp. 18–23.
[16] C.X. Mavromoustakis, K.G. Zerfiridis, On the diversity properties of wireless mobility with the user-centered temporal capacity awareness for EC in wireless devices, in: Proc. of 6th IEEE International Conference on Wireless and Mobile Communications, ICWMC 2010, Valencia, Spain, 2010, pp. 367–372.
[17] Y. Choi, H. Kim, S. Han, Y. Han, Joint resource allocation for parallel multi-radio access in heterogeneous wireless networks, IEEE Transactions on Wireless Communications 9 (11) (2010) 3324–3329.
[18] A. Bourdena, G. Mastorakis, E. Pallis, C.X. Mavromoustakis, G. Kormentzas, E. Karditsis, A radio resource management framework for opportunistic TVWS access, in: Proc. of 15th ACM International Conference on Modeling, Analysis and Simulation of Wireless and Mobile Systems (MSWiM' 12), Paphos, Cyprus, 2012, pp. 33–38.
[19] G. Mastorakis, C.X. Mavromoustakis, A. Bourdena, G. Kormentzas, E. Pallis, Maximizing energy conservation in a centralised cognitive radio network architecture, in: IEEE 18th International Workshop on Computer Aided Modeling and Design of Communication Links and Networks (CAMAD 2013), Berlin, Germany, 2013, pp. 175–179.
[20] C. Dimitriou, C.X. Mavromoustakis, G. Mastorakis, E. Pallis, On the performance response of delay-bounded energy-aware bandwidth allocation scheme in wireless networks, in: IEEE ICC2013, Budapest, Hungary, 2013.
[21] S.S. Skiena, The Algorithm Design Manual, Springer, 2011.
[22] C.X. Mavromoustakis, On the impact of caching and a model for storage-capacity measurements for energy conservation in asymmetrical wireless devices, in: IEEE Communication Society (COMSOC), 16th International Conference on Software, Telecommunications and Computer Networks (SoftCOM 2008), 2008, pp. 243–247.

[23] O. Diallo, J.P.C. Rodrigues, M. Sene, J. Lloret, Simulation framework for real-time database on WSNs, Journal of Network and Computer Applications 39 (2014) 191–201.
[24] COGEU-ICT-FP7-248560, COGnitive radio systems for efficient sharing of TV white spaces in EU context, http://www.ict-cogeu.eu.

Part III

# Commercialisation and Applications of White Space Networks

Chapter 7

# TV White Space Network Trials

**Juha Kalliovaara*, Jarkko Paavola†, Reijo Ekman†, Arto Kivinen† and Pekka Talmola‡**

**Technology Research Center, University of Turku, Turku, Finland, †Turku University of Applied Sciences, Turku, Finland, ‡Nokia, Espoo, Finland*

## 7.1 INTRODUCTION

This chapter describes TV White Space (TVWS) network measurements and trials conducted in the WISE (White Space Test Environment for Broadcast Frequencies) and WISE2 projects in Finland between 2011 and 2015. The TV white space test network environment was developed and built to aid standardisation work, and to demonstrate to the industry the technological capabilities of TVWS. A project consortium including universities, geolocation database providers, regulators, digital terrestrial television (DTT) broadcasters and user equipment manufacturers began to work on the WISE project [1] in 2011. The objective of the project was to construct a testbed to study the use of cognitive radios utilising white spaces in the UHF broadcasting band (470–790 MHz), to follow closely the topics in CEPT/ECC SE43 and to contribute the results obtained in the project within SE43. The first phase of the WISE project (2011–2013) focused on the interference measurements to protect the incumbents (Digital terrestrial TV and PMSE users) within the DTT band, while the second phase (WISE2, 2013–2015) focused on the application pilot trials.

The WISE project partners in the first phase were University of Turku, Turku University of Applied Sciences, Aalto University, Nokia, Digita, Fairspectrum and the Finnish Communications Regulatory Authority (FICORA). Additional industrial partners were introduced in the second phase: Nokia Siemens Networks, Elektrobit, Teleste, Suomen Turvakamera Oy, Satel, Helsingin Seudun Liikenne, City of Jyväskylä (Local traffic), Viola Systems, Insta DefSec and QEM Software. Fig. 7.1 illustrates the consortium partners in the first phase (inner circle) and in the second phase (both circles).

Section 7.1 gives a brief introduction to the TVWS trials conducted and the European level standardisation work, which was the main reason for building a test network environment for TVWS. Section 7.2 presents the TVWS test network environment, related radio licence, geolocation database and its radio

TV White Space Communications and Networks. http://dx.doi.org/10.1016/B978-0-08-100611-5.00010-2

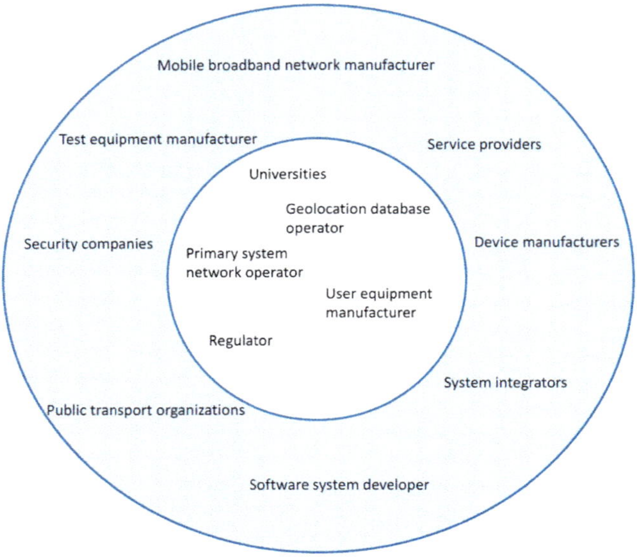


**FIGURE 7.1** Original WISE consortium in the middle and extended consortium of WISE2 at the outer circle.

environment mapping techniques, and the device emission measurements conducted for the white space devices (WSDs) used in the application pilot trials. Section 7.3 describes interference measurements conducted to study the protection of incumbents in the scenarios under consideration by the CEPT/ECC SE43 project team. Section 7.4 describes a video surveillance application pilot trial and the application pilot trials conducted with a public transport authority. Finally, the concluding remarks are given in Section 7.5.

### 7.1.1 Technical Work in the CEPT/ECC SE43 Group

Following a request from the Electronic Communications Committee (ECC), the European Conference of Postal and Telecommunications Administrations (CEPT) within the Working Group Spectrum Engineering (WG SE) established a new project team, Spectrum Engineering 43 (SE43), at its 53rd meeting in May 2009. The SE43 team began its work with the document 'Technical and operational requirements for the possible operation of cognitive radio systems in the "White Space" of the frequency band 470–790 MHz' [2]. SE43 met for the first time in June 2009, and completed the work at its 16th meeting in December 2012. All stakeholders were involved in the SE43 work: administrators, industry, operators, Programme Making and Special Events (PMSE) representatives and broadcasters.

SE43 was mandated to do the following:

- To define technical and operational requirements for the operation of cognitive radio systems in the white spaces of the UHF broadcasting band (470–790 MHz) to ensure the protection of incumbent radio services/systems and to investigate the consequential amount of spectrum potentially available as white space;
- To provide, if required, technical assistance on further issues related to white spaces and cognitive radio systems that the ECC may identify in the future;
- To liaise directly with relevant groups within the ECC and the European Telecommunications Standards Institute (ETSI) as necessary [3].

The outcome of the first phase of the work was ECC report 159 (Technical and operational requirements for the possible operation of cognitive radio systems in the white spaces of the frequency band 470–790 MHz) [4], which was released in January 2011. The report still had a number of technical and regulatory issues requiring further consideration, but it was decided that instead of updating it, two complementary reports would be released. The interference measurements conducted in this report addressed the issues mentioned in ECC report 159, and contributed to the second phase of the SE43 work, which resulted in two complementary reports: ECC report 185 (Complementary Report to ECC Report 159: Further definition of technical and operational requirements for the operation of white space devices in the band 470–790 MHz) [5] and ECC Report 186 (Technical and operational requirements for the operation of white space devices under geo-location approach) [6], which were released in January 2013.

The reports produced in SE43, and ECC reports in general, are studies in support of a harmonisation measure and have no direct regulatory power. Their importance is nevertheless extremely significant as they are used to aid European regulation and standardisation. The European Telecommunications Standards Institute (ETSI) is the official European standards organisation, which works under mandates from the European Commission to prepare harmonised standards under the provisions of the Radio and Telecommunication Terminal Equipment (R&TTE) Directive [7]. ETSI Broadband Radio Access Networks (BRAN) has used the reports produced by SE43 as the basis for its document 'Harmonised European Standard EN 301 598: White Space Devices (WSDs); Wireless Access Systems operating in the 470 MHz to 790 MHz TV broadcast band; Harmonised Standard covering the essential requirements of article 3.2 of the R&TTE Directive' [8]. This harmonised standard includes radio frequency requirements to prevent harmful interference from WSDs by ensuring that the wanted radiated power and unwanted radiated power do not exceed specified limits, which were set by using the results of the studies in SE43. The first phase of the WISE project contributed to the creation of this standard, while in the second phase of the project the standard was used to measure whether the WSDs used in the application pilot trials complied with the harmonised standard (see section 7.2.3). In June 2016, R&TTE was replaced by the Radio Equipment Directive (RED) [9], and the harmonised standards are currently being revised to meet the updated requirements of RED.

## 7.2 WISE – WHITE SPACE TEST ENVIRONMENT FOR BROADCAST FREQUENCIES

The Turku TV white space test environment was set up under the WISE project [1] to develop and validate technical solutions, accelerate the commercial utilisation of white spaces and to support the contributions to the regulation and standardisation work in the CEPT/ECC SE43 group. The test network and radio laboratory are located in Turku, Finland. TV white space equipment has also been installed and tested in different locations in Helsinki for use-case piloting.

The TVWS test network environment consists of the following components:

- A commercial level DVB-T terrestrial TV broadcast test network with three transmitters. The test network makes it possible to study the co-channel and adjacent channel interference of cognitive radios to live TV broadcast;
- A complementary simulation environment developed by University of Turku and Aalto University;
- A full cognitive radio licence for the TVWS frequency range 470 MHz–790 MHz;
- A Fairspectrum geolocation database;
- The Turku University of Applied Sciences radio laboratory, which is equipped with white space radios, spectrum analysis and spectrum measurement equipment.

FICORA has issued a test radio licence for cognitive radio devices in the TV White Space frequencies for Turku University of Applied Sciences. The licence covers the 470–790 MHz frequency range and a 40 km × 40 km area surrounding Turku, Finland. This area is indicated by a rectangle in Fig. 7.2. Nearly 300000 people live in the radio licence area. The licence is the first in Europe having a geolocation database in control of the frequency use [10].

Radio licence conditions are the following:

- Area: Turku region, 40 km × 40 km range (Fig. 7.2);
- Max +44 dBm EIRP for fixed station/base station;
- Max +30 dBm EIRP for mobile station/terminal;
- Maximum transmitter power: +31 dBm;
- Antenna gain 0–14 dBi (omni or directional antenna);

**FIGURE 7.2** Radio licence area.

- Antenna height: Max 40 m above sea level;
- Polarisation: vertical;
- Channel width: 6 MHz or 8 MHz;
- Number of transmitters: 1 Neul base station and terminal, 1 Carlson Wireless base station and 5 terminals;
- Transmission frequencies and power of the white space devices are controlled by a geolocation database. The database makes its decisions based on the location of the TVWS radio transmitters and the information on the primary users to be protected.

Fairspectrum provides the geolocation database for the TVWS test network according to the FICORA licence rules. The TV White Space radios of the test network were from the US manufacturer Carlson Wireless. One base station and five terminals were at our disposal. They operate in 470–786 MHz frequency bands, and both US 6 MHz and ETSI 8 MHz channel spacing can be used [11]. A prototype base station and terminal from UK-based Neul are also at our disposal.

The test network environment offers full control and flexibility in defining transmission parameters of the DVB-T network. Support for finding test locations and routes, estimating white space availability and other relevant information is available through simulations, calculations and signal level field measurements. Computer simulations can be used to verify and interpolate measurement results.

The test environment enables a wide range of testing possibilities in diverse environments, such as

- Service piloting;
- Interference management, protection ratios and co-existence;
- Algorithm development and optimisation.

### 7.2.1 Geolocation and PMSE Databases

The radio licence requires that the incumbent users of the UHF TV band need to be protected from any harmful interference. A geolocation database is used to provide protection for DTT and PMSE incumbents operating in the band from interfering white space devices. In practice, the geolocation database contains coverage maps of the incumbent DTT networks. These coverage maps are used to calculate the maximum allowed transmission powers for interference-free operation of the WSDs at their current geographical locations.

The incumbent PMSE users operating in the UHF TV band are wireless microphones. Turku University of Applied Sciences has developed a wireless microphone database to collect their operating parameters and locations to allow their protection. This database provides data to the geolocation database, which is then able to calculate the needed protection for both DTT and PMSE users. The protection zones are defined as geographical areas, which are then converted to polygons. The geolocation database utilises these polygons in the calculation of available channels and their associated power levels for different geographical locations. Geometrical computation makes the geolocation database efficient, fast and easily customisable for various environments, regulations and rules.

Before it can begin to transmit on TVWS frequencies, a WSD needs to connect to a geolocation database and to request available channels and their power levels by sending certain properties (antenna height, operational location, device type and device emission class) and transmission parameters. The geolocation database calculates the WSD's radio coverage area and analyses the DTT and PMSE transmissions within that coverage area, and finally provides the WSD with a channel to operate in and the maximum allowed power level in that channel.

The geolocation database can also be accessed using a web interface or mobile client software, which is illustrated in Fig. 7.3. Both allow the user to select a location from a map and to see the location's channel availability information calculated by the geolocation database. The channel information shows the availability and the maximum transmission power for WSD devices in each TV channel. Channels listed in grey are not available for WSD use. Web and mobile

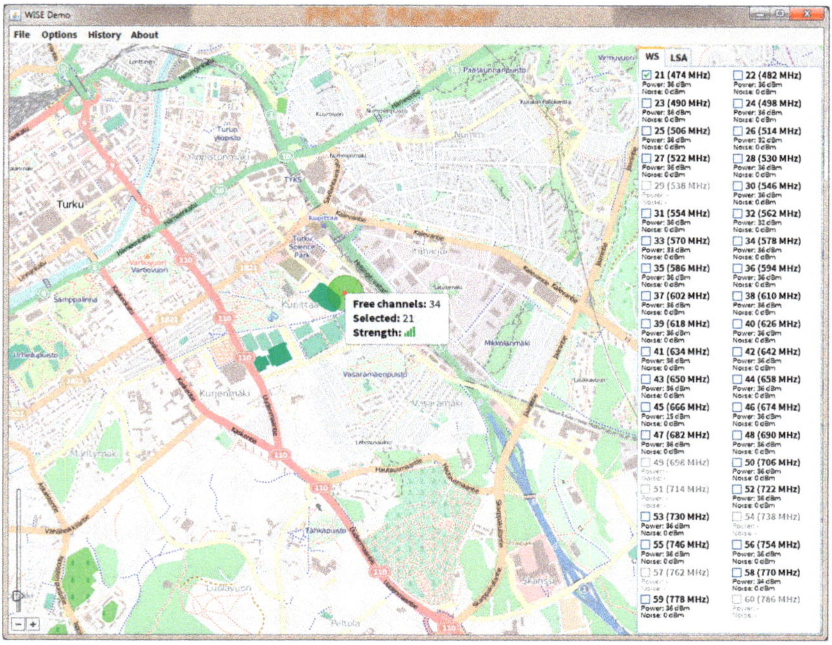


**FIGURE 7.3** Geolocation database client software.

clients use Protocol to Access White-Space Databases (PAWS) [12] to communicate with the geolocation database.

The PMSE Incumbent Manager was developed as a cloud-based web service where the users can add their PMSE devices to the PMSE database. The PMSE Incumbent Manager allows the users to add their wireless microphones by setting their location on a map and selecting a channel from the list of available channels. Time, location and frequency of a wireless microphone are needed by the PMSE database to enable the protection level to be calculated. The user first selects the location of the wireless microphone on a map; the PMSE Incumbent Manager then requests the channel and time-use information. This information is then communicated to the geolocation database.

### 7.2.2 Radio Environment Mapping Algorithms for the Geolocation Database

The key factor in TVWS operation is to calculate accurate maximum allowable power limits for WSDs. Accurate limits guarantee that the interference with the primary network remains within acceptable limits, while it also maximises the throughput of the TVWS system. Excessively strict power levels limit the

TVWS throughput, while limits that are too lenient cause harmful interference with the incumbents.

Radio environment mapping (REM) presents a more accurate alternative for signal level calculations than propagation modelling. Here we introduce hybrids of measurement campaigns and computer simulations which have been used to validate the signal levels of the DTT network in the geolocation database, and thus also the validity of the incumbent protection provided by the geolocation database.

In the creation of a radio environment map, measurements are conducted at some locations, and interpolation techniques are used to estimate values at locations which have not been measured. A sampling scheme makes decisions on where the samples are obtained from, i.e. the measurement locations. Since field measurements are resource-consuming, it is important to design a sampling scheme which allows an optimal estimation (interpolation) of the radio environment with a given number of measurement locations.

To overcome the limitations arising from using only radio propagation models or suboptimal field measurement locations, we have proposed a geostatistical modelling procedure for estimating the radio environment in [13]. An optimal sampling scheme was calculated using spatial simulated annealing (SSA) to minimise the mean universal kriging variance (MUKV) needed in the Universal kriging (UK) interpolation.

Fig. 7.4 shows an estimation based on a propagation prediction model in (A), and in (B) and (C) the estimation is performed by interpolating the minimum mean shortest distance (MMSD) scheme using inverse distance weighted interpolation (IDW) and ordinary kriging (OK) methods. These are earlier methods of calculation and interpolation to which the novel contribution in (D) is compared. In (D) the sampling locations are found by minimising the MUKV with SSA, and the field strength estimates for unsampled locations are determined by UK.

Field measurements were performed in locations indicated by grey crosses. The SSA algorithm was able to find the optimal measurement locations in 200 iterations. The measurement locations are strictly limited to the street network within the test area, and the modelling procedure was verified with an extensive measurement campaign performed in an operational DTT test network in Turku. A pixel size of 5 m $\times$ 5 m was used for this 2010 m $\times$ 2085 m area of the test network.

The proposed method of [13] presented in panel (D) of Fig. 7.4 was the most accurate according to the metrics used: mean absolute error (MAE), normalised root mean square error (NRMSE) and the Pearson correlation coefficient. It significantly improved the precision of local field strength estimates compared to

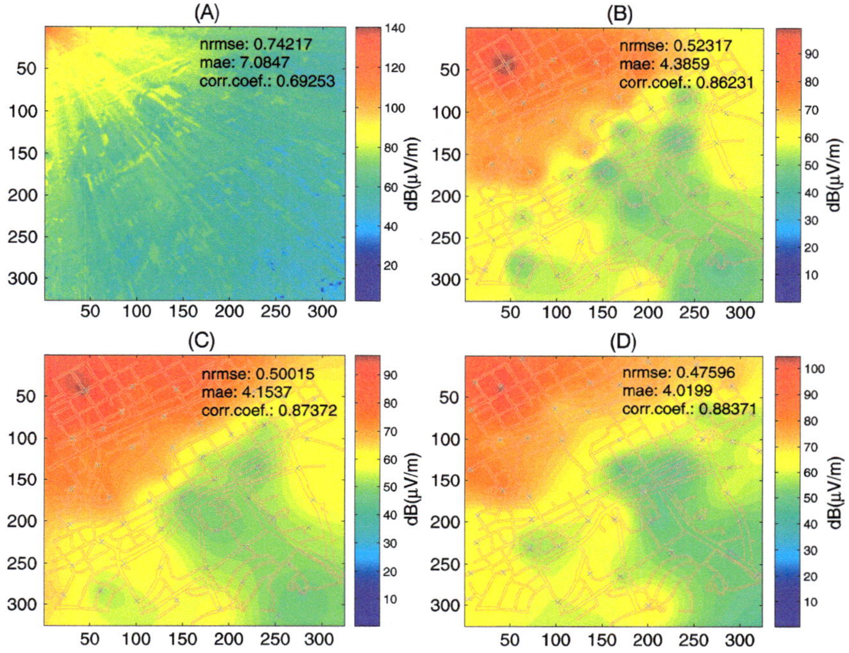


**FIGURE 7.4** Estimated field strength values for the Turku DTT test network area with the methods used in [13]. In (A) a prediction is presented based on a propagation model. In (B) and (C) the field strength is estimated by interpolating the samples obtained from locations chosen with MMSD via IDW and OK methods. The sampling locations in (D) are selected by minimising the MUKV, and the field strengths are estimated with UK interpolation. Grey crosses indicate the locations where the field measurements have been performed.

using only propagation model (A), or by interpolating the field measurement values sampled on a regular grid (B and C).

In [14], we have designed an efficient technique for improving the REM estimation in scenarios where relatively small numbers of measurement samples are available. The proposed multivariate kriging method utilises correlated secondary information obtained from a terrain-based propagation prediction model [15] to complement the measurement data. Considerable improvement in prediction accuracy is achieved compared to univariate interpolation methods.

The proposed techniques can be utilised in designing sensing networks or in measurement campaigns for REM. The radio environment maps can be used in the TV white space geolocation databases to calculate accurate protection levels for the DTT networks. The accurate signal level estimations also help when searching for locations with specific signal levels within the test network, such as in outdoor reference geometry measurements (see Section 7.3.1.1).

### 7.2.3 Device Emission Class Measurements

Prior to performing trials in the TVWS test network, the unwanted emission levels of a WSD need to be verified. ETSI EN 301 598 harmonised standard for white space devices [8] is used for this purpose. This standard is based on the SE43 work, as described in section 7.1.1. The standard defines five different device emission classes for WSDs. This class is an important parameter in the geolocation database, and it is used in the calculation of allowed transmitter power level for the WSDs. A correct classification is essential to avoid harmful interference to incumbents of the UHF TV band.

Turku University of Applied Sciences (TUAS) Radio laboratory has a measurement system to conduct unwanted emission measurements to define and verify the device emission class according to the ETSI EN 301 598 [8]. The WSD adjacent channel leakage ratios (ACLRs) are measured within the 470 to 790 MHz TVWS operational band, and the emissions outside the band are measured from 30 MHz to 4 GHz. Table 1 of [8] presents the requirements for Adjacent Channel Leakage Ratio (ACLR) for different emission classes, and Table 2 of [8] the requirements for transmitter unwanted emission limits outside the 470 to 790 MHz band.

The measurement requirements for pre-scan sweeps are resolution bandwidth of 100 kHz in 30–1000 MHz and 1 MHz in 1–4 GHz frequency bands, peak detection and trace mode max hold. Emissions which are above the limits in the pre-scan are individually measured in a frequency band wide enough to capture each identified emission using continuous sweep, root mean square (RMS) detector and max hold trace mode.

The better the classification of the device under test (DUT) is, the less out of band (OOB) emissions there are and the more power can be granted by the geolocation database for WSD operation. Five classes are specified, with class 1 representing WSDs with the lowest levels of unwanted emissions and class 5 representing WSDs with the highest levels of unwanted emissions. The performance of the DUT is usually not constant in different channels. For example, a DUT can pass class 1 criteria on most of the channels, while only passing class 4 criteria on the other channels. The worst measurement result defines the classification of the DUT, so in this case the DUT would pass only as a class 4 device even though it performs better in most of the channels.

## 7.3 MEASUREMENTS CONTRIBUTED TO CEPT/ECC SE43

This section describes the measurements conducted during the WISE project between 2011 and 2013 to investigate the technical and regulatory issues designated as needing more consideration in ECC report 159. The DTT-WSD protection ratio measurements in section 7.3.1 have been contributed to CEPT/ECC SE43 in [16–19] and the PMSE measurements of section 7.3.2 in [20,21].

### 7.3.1 Protection Ratio Measurements in Reference Geometries Defined in ECC Report 159

The existing primary licensed users in the radio spectrum are also known as incumbents. Their transmissions need to be protected from harmful interference from secondary users, such as white space devices. Various methodologies have been used to develop protection criteria for the incumbents in the UHF TV band, including calculations with parameters from ECC, International Telecommunication Union Radiocommunication sector (ITU-R) and ETSI deliverables. Several measurement campaigns regarding the use of WSDs in TV frequencies were conducted before 2011, but these efforts were not able to clearly give safe operational rules for WSDs. This resulted in a number of different recommendations for the technical and operational requirements that should be applied to WSDs operating in the frequency band 470–790 MHz in order to protect the incumbent DTT transmissions and PMSE wireless microphones within the band. Technical work to define European-wide protection ratio proposals was thus started by the CEPT/ECC project team SE43.

The issues regarding the DTT incumbent protection were discussed in ECC report 159 [4] released in 2011. Two cases must be distinguished when considering the interference caused by a WSD to DTT reception. In the first case the WSD and the victim DTT receiver are separated by a relatively large distance and interference is primarily of co-channel type. The DTT receiver location can be estimated as the nearest possible pixel of the DTT service area in question, and the coupling loss between the WSD and DTT receiver can be calculated based on their distance and a suitable propagation model. In the second case the WSD and the victim DTT receiver are within the same pixel. In this case the interference is typically of adjacent channel type, as co-channel operation is not permitted within the DTT service area. Calculating the coupling loss in this case is problematic as nothing is known about the geometry between the WSD and the DTT receiver, only that they may exist within the same pixel and thus their distance is constrained by the pixel dimensions. Several reference geometries were proposed to simulate different worst-case DTT reception scenarios. If the DTT reception is protected in a worst-case scenario, it is assumed to be protected in any other scenario as well.

An indoor and an outdoor DTT reception reference geometry from ECC report 159 were chosen to be used in WISE project field measurements to study protection ratios between WSDs and DTT.

In [22] a protection ratio (PR) is defined as the minimum value of the signal-to-interference ratio required to obtain a specific reception quality at the receiver input. Formally, this ratio is defined as

$$PR(f_{pri}, f_{wsd}) = \frac{P_{pri}(f_{pri})}{P_{wsd,max}(f_{wsd})}, \tag{7.1}$$

that is, the ratio of the received TV signal power centred at frequency $f_{pri}$ to the maximum allowable power transmitted by the WSD at centre frequency $f_{wsd}$. In these measurements, the maximum WSD transmit power is determined as the limit where error-free TV reception is still possible according to a subjective criterion corresponding approximately to the ESR5 (Erroneous Second Ratio 5%) criterion [23].

The presented measurement campaigns can be considered as the first practical testing of the proposed theoretical reference geometry WSD scenarios, but also as a source of realistic numerical estimates for the minimum protection ratios between WSD transmitters and TV receivers; as such, the given data can be used for example in further WSD system development, simulation and performance evaluation. The results have been contributed to the SE43 group meetings in [16–19] and published in the proceedings of 7th International Conference on Cognitive Radio Oriented Wireless Networks (CROWNCOM 2012, Stockholm, Sweden, June 18–20, 2012) [24].

#### *7.3.1.1 Field Measurements in Outdoor Reference Geometry*

The outdoor reference geometry defined in ECC report 159 and used in the field measurements is for rooftop antenna DTT reception at a height of 10 m with a portable WSD at 1.5 m and 22 m from the DTT reception antenna. An omnidirectional antenna is assumed for the WSD so that a polarisation discrimination of 3 dB is possible, but in the measurements a directional antenna with horizontal polarisation for the WSD was also used to increase the available WSD equivalent isotropically radiated power (EIRP). The DTT antenna specified in the reference geometry has a gain of +9.15 dBi, which is assumed to drop by 0.45 dB in the direction of the WSD 22 m away. The antenna is described in ITU-R BT 1368 [25]. This geometry aims to represent a worst-case scenario: if the WSD is situated closer the antenna gain will drop, and if the WSD is further away the free space loss will increase, decreasing the interfering signal power at the DTT receiver. The considered outdoor reference geometry is illustrated in Fig. 7.5.

The outdoor measurements study the maximum possible WSD power in the corresponding reference geometry before the received picture fails. The measurements were done at three different DTT signal levels. The first was at an input level of −80 dBm, 3 dB from the sensitivity limit of the receiver, corresponding to the situation where the DTT receiver is located at the edge of the

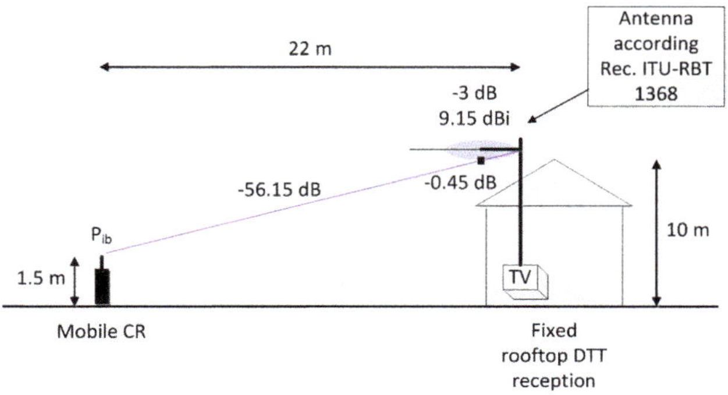


**FIGURE 7.5** Outdoor reference geometry.

**FIGURE 7.6** TVWS radio antenna at 1.5 m above ground level (on the right) and 10 m DTT reception antenna (on the left) in an outdoor reference geometry measurement.

service area. The second was with an input level of −70 dBm and the third with −62 dBm input level at the receiver. The locations of the measurements were selected so that a suitable signal level was achieved with a Yagi antenna at a height of 10 m. One field measurement location and measurement antenna set-up is shown in Fig. 7.6.

Interference measurements were performed with the WSD operating on the co-channel $N$ and on adjacent channels $N-1$, $N-2$, $N-3$, $N-4$ and $N+9$. The WSD transmission power was increased until errors were detected in the picture and then decreased by 1 dB steps until no errors were visible during an observation period of several tens of seconds, corresponding roughly to the ESR5 criterion [23]. Then the power fed to the WSD transmission antenna was measured in the co-channel case. In the adjacent channel cases, the received

**TABLE 7.1** Outdoor Protection Ratios With 64-QAM DVB-T

| Frequency (MHz)/(Channel) | DVB-T level −80 dBm | | | DVB-T level −70 dBm | | | DVB-T level −62 dBm | | |
|---|---|---|---|---|---|---|---|---|---|
| | WSD | | PR | WSD | | PR | WSD | | PR |
| | EIRP (dBm) | FS (dbμV/m) | (dB) | EIRP (dBm) | FS (dbμV/m) | (dB) | EIRP (dBm) | FS (dbμV/m) | (dB) |
| 578/($N-4$) | 18.9 | 94.0 | 48.4 | 32.5 | 107.6 | 51.9 | 36.4 | 111.5 | 49.0 |
| 586/($N-3$) | 18.3 | 93.4 | 47.5 | 21.5 | 96.6 | 40.6 | 23.0 | 98.1 | 35.4 |
| 594/($N-2$) | 15.2 | 90.2 | 44.1 | 18.7 | 93.7 | 38.4 | 20.5 | 95.5 | 33.2 |
| 602/($N-1$) | 1.7 | 77.0 | 31.4 | 13.3 | 88.6 | 33.2 | 12.8 | 88.1 | 25.9 |
| 610/($N$) | −50.0 | 24.7 | −19.7 | −36.5 | 38.2 | −16.2 | −30.0 | 44.7 | −17.7 |
| 682/($N+9$) | 20.9 | 95.6 | 51.1 | 33.8 | 108.5 | 53.7 | 38.5 | 113.2 | **48.4** |

WSD interference power at the DTT receiver input was also measured and used to calculate the corresponding protection ratio. It should be noted that with the WSD operating on the adjacent channel $N+9$ in the scenario with DTT signal strength −62 dBm, the WSD transmission power was insufficient to produce errors in the DTT reception, and the true protection ratio is higher than given here. This case is marked in bold text in Table 7.1, which shows a summary of the maximum WSD power levels for different frequencies and DTT input levels. The corresponding Field Strength (FS) at the DVB-T reception site is also shown, as is the protection ratio calculated with formula (7.1). The ACLR values for each measurement and the laboratory measurements of the DTT receivers can be found in [17].

### *7.3.1.2 Field Measurements in Indoor Reference Geometry*

The portable indoor reference geometry defined in ECC report 159 and used in these measurements is shown in Fig. 7.7. Both the DTT reception antenna and the WSD transmitter are at a height of 1.5 m and are 2 m from each other. Omnidirectional antennas are assumed for both the WSD and DTT with no po-

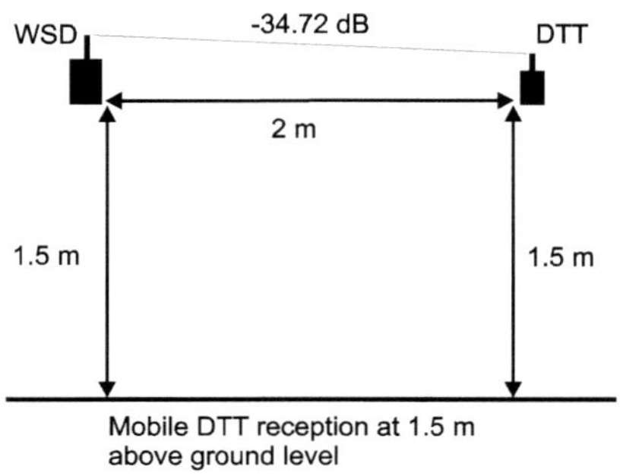


**FIGURE 7.7** Indoor reference geometry.

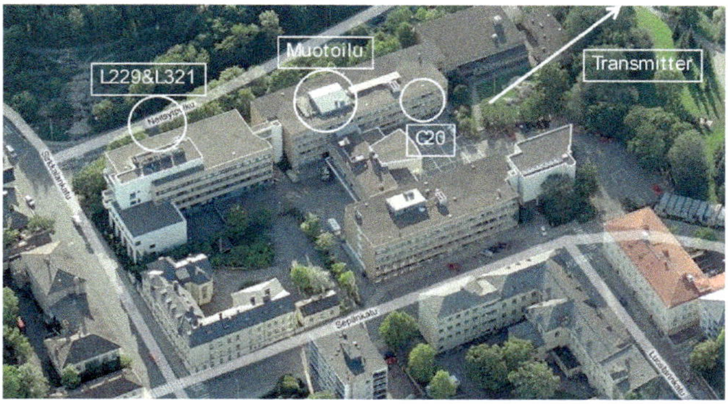


**FIGURE 7.8** Indoor measurement premises.

larisation discrimination. This was achieved in the measurements by using two similar omnidirectional antennas with a nominal gain of 2 dBi.

All indoor measurements were performed in an old school building constructed mainly of brick and concrete, and whose windows had no metallised shielding. The building, measurement locations, and transmitter direction are illustrated in Fig. 7.8. It should be noted that DTT transmissions penetrate this type of building better than a modern office building. The following rooms were used for the measurements:

- 'Muotoilutila' (Muo): a small rooftop space used for project works;
- L229: classroom in L-wing, second floor;
- C20: classroom in C-wing, second floor;
- L321: classroom in L-wing, third floor.

The 'Muotoilutila' is a small room on top of the C-wing and has a window towards the transmitter. All the other rooms have windows perpendicular to the transmitter direction so no direct signal was received in these cases.

The rooms where the measurements were performed have different geometries and thus different reflections from the surroundings. Therefore, the coupling losses also differ between the rooms even though the distance between the WSD and DTT is constant. The coupling losses were measured in each room in the 2 m reference geometry. A comparison between these losses and a calculated free-space loss (FSL) is shown in Fig. 7.9. Even though the number of measurements is not large enough to give statistically reliable results, a clear trend in the measurements is that the coupling loss is higher than the FSL.

The indoor measurements again studied the maximum possible WSD power in the reference geometry before the DTT reception failed. Most of the measurements were made close to the sensitivity limit of the DTT receiver, corresponding to the situation where the DTT receiver is located at the edge of the

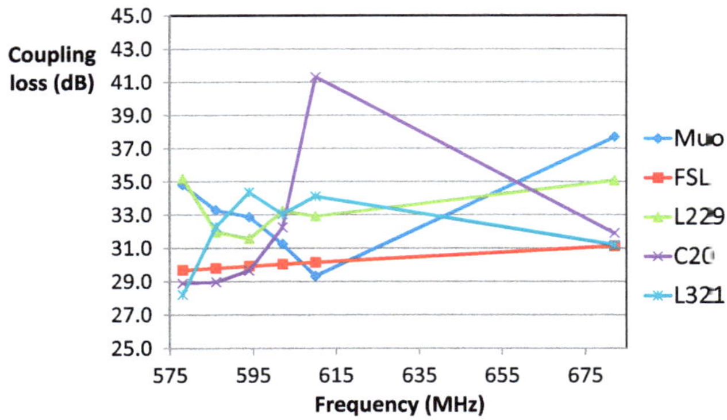


**FIGURE 7.9** Free-space loss and coupling losses in different rooms.

**TABLE 7.2 Indoor Protection Ratios With 256-QAM DVB-T2**

| Frequency (MHz)/(Channel) | Protection ratio with 32k 256-QAM 3/5 in room | | | | |
|---|---|---|---|---|---|
| | Muo 2m | L229 2m | C20 2m | L321 2m | Average |
| 578/($N-4$) | 43.0 | **40.1** | 43.9 | 44.5 | 42.9 |
| 586/($N-3$) | 41.9 | 42.5 | 42.9 | 40.1 | 41.9 |
| 594/($N-2$) | 39.5 | 40.8 | 40.9 | 38.9 | 40.0 |
| 602/($N-1$) | 25.6 | 28.6 | 35.0 | 30.7 | 30.0 |
| 610/($N$) | −23.5 | −23.6 | −22.8 | −23.8 | −23.4 |

service area or in otherwise difficult indoor conditions. The protection ratios were calculated from the measured WSD and DTT powers at the DTT receiver input. Results are shown in Table 7.2. In one measurement, L 229 2m/$N-4$, the available WSD power was not high enough to cause any errors, and thus the calculated protection ratio is likely to be lower than the real value would be. This case is marked in bold text in Table 7.2. The protection ratio values are quite consistent. The greatest spread is in the values at $N-1$ (602 MHz). On average, the co-channel protection ratio is between −20 and −25 dB, the protection ratio on channel $N-1$ is between 25 and 35 dB, on $N-2$ between 39 and 42 dB, and on $N-3$ and $N-4$ between 40 and 45 dB. The variations are at least partly due to different channel conditions at different locations. The laboratory measurements performed to calibrate the measurement set-up are explained in [18].

## 7.3.2 PMSE Wireless Microphone Measurements

PMSE is a term which covers equipment that supports broadcasting, news gathering, theatrical productions, sport events, etc. PMSE devices operating in the TV White Space frequencies tend mainly to be wireless microphones. Geolocation databases are currently considered to be the primary method of protecting PMSE users as sensing techniques are currently not sufficient to provide reliable protection for PMSE, and the range of potential deployment scenarios causes large variability in the sensing thresholds. In the following we outline a measurement campaign conducted to evaluate interference between WSD and PMSE devices in a real use scenario.

Two theatres in Helsinki were used for the measurement campaign. Helsinki City Theatre, located in Kallio, is the biggest theatre in Helsinki. The main building was built in the 1960s and is a traditional concrete and steel construction. It has two stages, a big stage with 947 seats and a smaller stage with 400 seats. The Arena Theatre is a smaller 515-seat theatre. Arena is located near the City Theatre in a larger brick building built in 1923. Illustrations of the two theatres are available in the full measurement report [20].

Two sets of analogue microphones were used in the measurements, altogether four microphones and two receivers. The purpose was to use microphones in real operating conditions, and therefore the receivers were placed at realistic places in the theatres. At the main stage of the City Theatre, the receiver was placed in the middle of a balcony above the audience. In the Arena theatre, the receivers were placed on the balcony as well, and close to the existing antenna installations in the theatre. In both arenas, the microphone to receiver link in the measurements was somewhat more demanding than with the existing installations. The microphone receivers used small whip antennas attached directly to the receivers.

During the measurements we found that there is a large difference in the results depending on how the microphones are used. The worst case, but also the most realistic, was the belt pack microphone attached to a person moving around the stage. The loss due to the bodies of the actors and the movement is at its highest in this scenario. Therefore, the most realistic scenarios were those where the microphones were attached to people moving around the stage and simulating real actors, sometimes even going behind the stage.

Interference to the PMSE equipment was caused by a simulated WSD operating on the co- or adjacent channel. The measurements were performed with the WSD in several locations inside and outside the building. Fig. 7.10 illustrates the locations of the microphones and PMSE receivers, and the WSD locations (yellow circles) inside the Helsinki City Theatre. Subjective evaluation was used to determine when the WSD was causing audible interference.

The measurements clearly show that a WSD operating co-channel with the PMSE causes interference on very low power levels ($-15$ dBm ... $+5$ dBm) when the WSD is in the vicinity of the PMSE equipment. Thus, co-channel operation of PMSE equipment and WSDs is not possible in the vicinity of PMSE receivers. The protection ratios on the adjacent channel were approximately

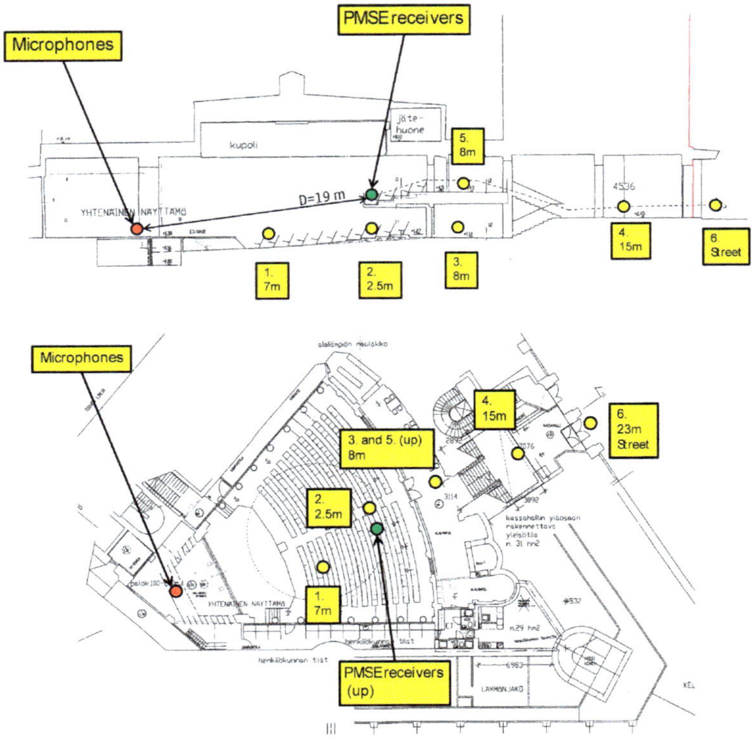


**FIGURE 7.10** Locations of the PMSE equipment and WSD transmitters measured at Helsinki City Theatre. Figure adopted from [20].

+30 dBm and on the next adjacent channels +40 dBm or more. These values would allow adjacent channel operation with reasonable power levels for the WSDs located close to the PMSE equipment. In the measurements conducted at Helsinki City Theatre it was not possible to cause co-channel interference at a distance of 560 m with the maximum transmit power of approximately 10 W.

The results of the measurement campaign were presented at the 12th SE43 meeting in Cambridge, the United Kingdom, in December 2011 [20,21]. The key findings were included in ECC report 185 [5] and the full measurement report added as an annex to the report.

## 7.4 APPLICATION PILOT TRIALS

This section presents two TVWS application pilot trials conducted during the second phase of the WISE project between 2013 and 2015. The first presented application pilot trial was a long-term video surveillance trial in the Turku

TVWS test network environment. The second was an application pilot trial with the public transport authority in Helsinki, where the TVWS devices were tested to provide connectivity for their ticket sales and information screen systems. The application pilot trials are covered in greater detail in these studies (published in Finnish as Bachelor's theses) organised by the Turku University of Applied Sciences: the video surveillance trial in [26,27] and the public transport trial in [28].

### 7.4.1 Video Surveillance Trial

The video surveillance application pilot trials were conducted utilising the TUAS radio laboratory commercial TVWS cognitive radio systems from Carlson [11], a cognitive radio licence from FICORA and the Fairspectrum geolocation database. The data transmitted over the TVWS radio link was in the form of video signal from surveillance cameras.

Carlson TVWS radio system performance measurements were first conducted to verify the performance data provided by the manufacturer. This data was used in the system deployment design and field trial planning. The Carlson system supports a 16 Mbps data rate, but it has asymmetric uplink/downlink speeds (30/70%). The system is designed to provide rural broadband service, where more capacity is typically needed in the downlink direction. In video surveillance applications this is not an optimal ratio as the video cameras are installed in the terminal sites and the video content delivery needs higher bit rates in the uplink than the camera control data in the downlink direction does. Thus, the asymmetry should be the other way around to provide optimal performance for the video surveillance application. The data rate in both directions depends on the modulation used and error correction coding. The system supports 16-QAM, QPSK and BPSK modulation with several different code rates. Maximum bit rates measured in the laboratory tests were 10 Mbps for downlink and 6 Mbps for uplink with 16-QAM modulation and 3/4 code rates.

Experimental research and several field measurement campaigns were conducted to investigate the interference between the DTT receivers and the Carlson TVWS radio systems (i.e. white space devices). Interference from the adjacent ($N+1$ and $N-1$) and alternate channels ($N+/-2, 3, 4, \ldots$) was measured using commercial DTT signals from the local broadcasting stations, and also from more distant DTT broadcasting stations.

In the first interference measurement campaign, the TVWS radio terminal antenna was mounted on the same mast with the DTT reception antenna, and separate antennas with distances of 2 to 5 metres were also used. Field strengths from the remote DTT transmitters were measured to study the effects on the TVWS radio base station uplink co-channel reception. The protection ratios be-

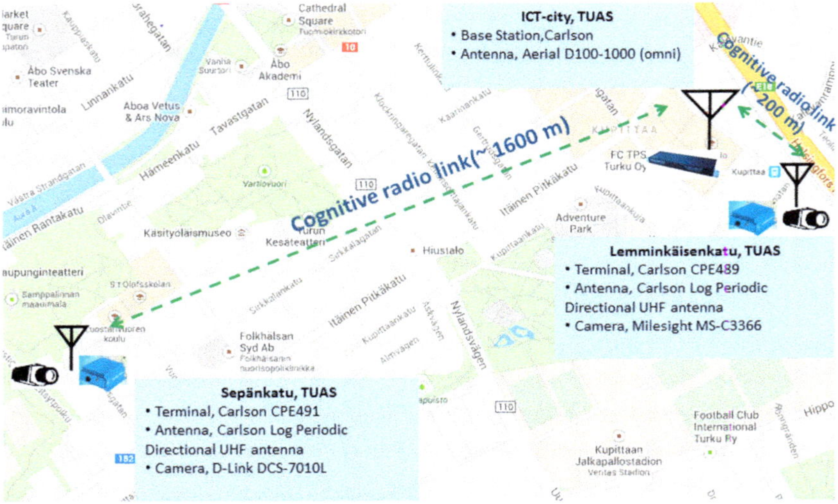


**FIGURE 7.11** Transmitter and base station locations of the TV white space wireless video surveillance system.

tween the TVWS radio and the DTT signal in the adjacent channel are in the order of 20–50 dB for commercial TVs and set-top boxes (STBs). The measurement data and results of this campaign were used in the verification and further development of the geolocation database.

### *7.4.1.1 TVWS Video Surveillance System Set-Up*

The video surveillance application was designed and implemented with a Carlson TVWS radio system and two commercial standard definition (SD) and high definition (HD) video cameras. The base station and terminals were installed within the Turku TVWS test network licence area. The base station was located on a rooftop of the ICT-city building and the camera sites were at Sepänkatu with a distance of 1.6 km and at Lemminkäisenkatu with a distance of 200 metres. The cameras and their related equipment in the sites were installed on rooftops at 15–20 metres above ground level (AGL) and the base station antenna on a rooftop at a height of 30 metres AGL. The base station antenna is omnidirectional, and the terminal antennas are log-periodic directional. The TVWS radio system terminal access is based on time division multiplexing (TDM), where the terminals share the payload time slots depending on the data rate required for the service. Hence, the two cameras share the time slots between the short control transmissions of the base station. The locations of the terminals and the base station are illustrated in Fig. 7.11.

The video camera and TVWS radio settings were as follows:

- Site 1/200 m link. Camera: resolution 1920 × 1080 at 25 frames per second (fps). Terminal: modulation 16-QAM with a code rate of 3/4.
- Site 2/1.6 km link. Camera: resolution 1280 × 720 at 25 fps. Terminal: modulation QPSK with a code rate of 3/4.

Measured signal-to-noise ratio (S/N) in the Site 1 link was 27 dB, which supports the use of high modulations. The site 1 camera was pointed towards a motorway with almost constant traffic and movement in the picture. The video setting Variable Bit Rate (VBR) gave better quality than Constant Bit Rate (CBR). A bit rate of 3 Mbps was sufficient for smooth video.

On Site 2, the measured S/N was 15–20 dB. The site 2 antenna was pointing over the roofs of nearby buildings, resulting in stationary picture content and minor movements. The CBR setting and 2 Mbps provided sufficient picture quality in this environment.

The maximum uplink bit rates at the sites were the mentioned 3 Mbps and 2 Mbps, as only 30% of the bandwidth was used for uplink due to the asymmetry in the time-division multiplexing of the TVWS radio system. It was possible to implement HD video surveillance with the Carlson TVWS cognitive radio system, but the 70%/30% downlink/uplink ratio is far from optimal. A ratio of 10%/90% could support 12–13 Mbps capacity for cameras. Depending on the video parameters and bit rates, this could allow installation of 5–6 cameras and terminals to be used with one base station.

#### *7.4.1.2 Long-Term Measurements for the TVWS Video Surveillance System*

Long-term field measurements were conducted to investigate the reliability of the TVWS cognitive radio link and the service quality in real-time video surveillance. The TVWS radio system is connected to the geolocation database through the Internet. The database controls the TVWS radio system and can evacuate the TVWS radio transmissions if the primary user needs the frequency resources which are utilised by the TVWS radio. TVWS radio channel S/N could be seriously decreased by remote DTT transmissions, which can propagate over long distances in certain weather and atmospheric conditions. Man-made noise can also cause similar effects.

The TVWS radio system downlink and uplink traffic information is stored by the system control manager software. The variation between daytime and night-time can be clearly noticed in the traffic. The data rate at night-time is very low, as it is very dark outside and the camera is directed towards a dark horizon. The downlink traffic is always very low, but the variation between day and night is also present in this direction. When the video monitoring software

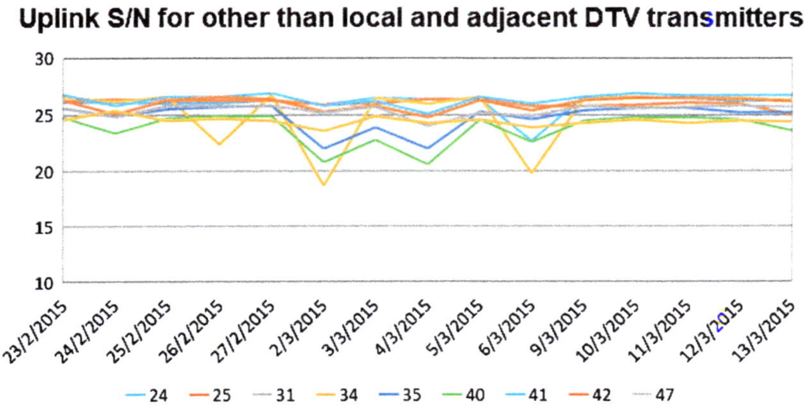


**FIGURE 7.12** Uplink S/N (dB) for allowed UHF TV channels.

is set to recording mode, it tries to minimise the amount of stored data and the variation between day and night is present. If the recording mode is not in use, the data rate is constant throughout the whole day. Uplink S/N for all the allowed channels for TVWS operation in the UHF band was measured and recorded between 23 February 2015 and 13 March 2015. Fig. 7.12 illustrates that a decrease in the channel S/N can be seen in three of these channels. The highest decrease is 7 dB in channel 31, but channels 34 and 35 also decreased by 2–5 dB over three consecutive days. The base station antenna is omnidirectional and located at a greater height than the directional antennas of the terminals. This causes higher received DTT signal levels at the base station antenna, which result in a stronger decrease in the signal quality in the uplink direction from the terminals.

## 7.4.2 Public Transport Application Pilot Trial

The purpose of the public transport application pilot trial was to create radio links in white space frequencies in real operating conditions with real use cases. Helsinki Regional Transport Authority (in Finnish: Helsingin seudun liikenne (HSL)) is responsible for planning and procuring the public transportation in the greater Helsinki area with its 1.1 million inhabitants. The public transportation in the area includes bus, tram, metro, ferry and commuter train services.

HSL uses many different types of services that rely on wireless communications; two use cases were chosen as pilot trials to investigate the eligibility of TVWS radio links: ticket sales and real time transit information screens. Currently, both of the systems are implemented with 3G mobile broadband connections. The goal was to study if a TVWS system could replace these 3G communication systems. The main difference between the use cases is that ticket sales uses two-way communications, while the transit information screens only receive the information updates but do not communicate any information.

The higher the base station antenna mast of the TVWS system is, the higher the DTT channel signal levels will be at the antenna. These high DTT signal levels present a huge challenge for the low-power TVWS terminal uplink connectivity to the base station. However, this was a problem only in the ticket sales application, where two-way communications were used. The transit information screen application allowed the use of higher antenna installations for the TVWS base stations as no information was transmitted back to the base station from the TVWS terminals.

FICORA issued a test licence for the application pilot trial for the central Helsinki area. The licence was valid only during the measurement days and only for the Carlson TVWS cognitive radios [11], which were to be controlled by the geolocation database.

The properties of the equipment HSL used in their applications were tested in a laboratory environment before the field measurements. The TWVS radios were used as their communication method to investigate the connection requirements for each application. The ticket sales required a faster connection, which required the use of 16-QAM modulation with our TVWS radio. The information screens require very low data rates, and thus the necessary throughput could be achieved with BPSK modulation. With the knowledge obtained regarding the required modulations and their sensitivity levels, propagation models were then used to create initial estimations on the possible link distances that could be achieved with the TVWS radios in the two application pilots.

### *7.4.2.1 Ticket Sales Trial*

The base station antenna was installed on a rooftop at a height of 45 metres AGL in the ticket sales application. The terminal was used in the measurement locations with an antenna installed on a four-metre tripod. To achieve comparable results between different locations, channel 41 (634 MHz) was used for all of the measurements. Transmission powers of 23.2 dBm for downlink and 20.7 dBm for uplink were used. The interference level at the base station antenna was already high enough to cause interference for the uplink transmissions, as was expected with the combination of strong DTT signals and the antenna on a high rooftop. This resulted in very short link distances for the ticket sales application, which needed two-way communications with low latency.

The ticket sales application became extremely slow if modulations of lower order than 16-QAM were used, making them practically unusable. A 16-QAM uplink needed a line of sight to operate in the urban environments of Helsinki.

**FIGURE 7.13** Base station antenna on Pasila TV tower.

The information screens, which only require communications in the downlink direction and could be operated with lower order modulations, were also measured and tested in the locations-with more success. Still, these first-generation TVWS radios cannot be recommended for this application in an urban environment. The longest achieved radio link distance was only 1166 metres.

### *7.4.2.2 Transit Information Screen Trial*

Another measurement campaign focusing on the transit information screens was conducted to further analyse the propagation characteristics of UHF TV frequencies in a city environment. With the transit information screens, the base station antenna height was no longer problematic as no uplink connectivity was required. The base station antenna was installed on a TV tower with an antenna height of 99.1 m AGL (128.3 m above sea level). The base station antenna installation is shown in Fig. 7.13.

The base station had a vertically polarised panel antenna (Rymsa AT15-250) with a 26 degree main lobe directed to the south (169°). An antenna installed at this height picks up distant DTT transmissions at very high power levels. As can be seen from Fig. 7.14, the measured DTT signal levels from the antenna are so high that they make the uplink connectivity towards the base station very difficult. Vacant channel 24 (498 MHz) was used for all of the measurements conducted with this antenna installation.

The transit information screens where the measurements were conducted were installed at tram stops, which allowed us to build a portable measurement set-up and use the trams to move between the measurement locations. The tram stops for the measurements were chosen mainly within the main lobe of the base station antenna. In addition, two places in Pikku-Musta and Iso Mustasaari islands were measured, as well as the signal level on the ferry from the mainland

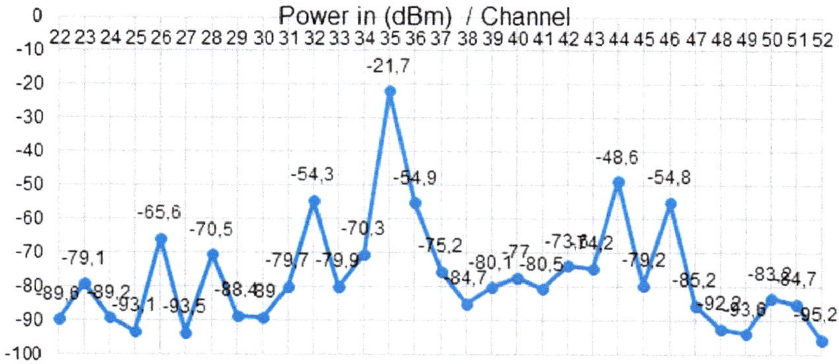


**FIGURE 7.14** Channel powers in Pasila TV tower (Antenna gain 12 dBi, vertical polarisation).

to these islands. Fig. 7.15 shows an overview of the measurement results on a map. All of the locations with green signal levels (higher than −70 dBm) in central Helsinki were measured in places with a lot of open space or at higher elevations, e.g. hills. Even in the islands at 7 km from the base station, the signal levels were significantly better than in the city centre at 3–5 km from the base station. The islands had a lot of open space and line-of-sight conditions towards the base station antenna, while central Helsinki is full of tall buildings and thus non-line-of-sight conditions. The locations close to the open space of Töölönlahti Bay and up on the higher elevations of the Observatory hills were the only locations with consistent signal levels and operating conditions in central Helsinki.

Thus, the environment proved to be too challenging for the TVWS radios used. However, in the islands even the uplink worked well. The TVWS radios used in the measurements are not built for urban environments, which could be seen from the measurement results. In the rural-like conditions of the islands they performed well. Point-to-point operation with a line of sight provides good connection with these radios in rural conditions, but they cannot be recommended for non-line-of-sight operating conditions in urban environments required by these application pilots.

## 7.5 CONCLUSIONS

This chapter has presented a TV white space test network environment built in Turku, Finland, and its related geolocation database. The interference measurements conducted in the test network environment contributed to the standardisation work in CEPT/ECC SE43 project team, whose results and reports have been used to create an ETSI harmonised standard for TV white space devices. The geolocation database and its interference protection algorithms have been

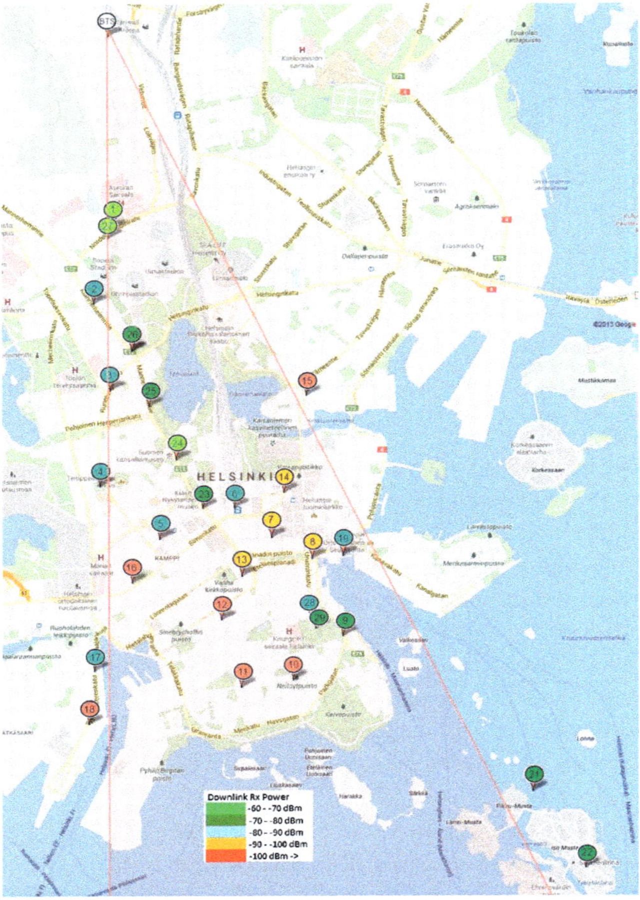


**FIGURE 7.15** Pasila TV tower signal levels at measurement locations.

developed and validated concurrently with realistic protection ratio information obtained from interference measurements in field conditions.

The technological capabilities of TV white space cognitive radios were demonstrated to industry through different application pilot trials between 2013 and 2015. The penetration and range in the TVWS frequencies are extremely good, but the equipment used in the application pilot trials was more suitable for fixed or rural installations than for mobile use or urban environments. This was expected as the units used were early first-generation TVWS radios specifically designed for rural use. Better performance in urban environments can be expected from the new generations of TVWS radios which are available from several manufacturers. The application pilot trials have shown that commercial deployment of TVWS radios is possible from the technological viewpoint, but that the limitations of the early generation TVWS radios need to be taken into account.

From technological and regulatory viewpoints there are no obstacles for the European national administrations to deploy TVWS networks. In the UK such networks are already operational, but it is yet to be seen what will happen regarding TVWS network deployments in Finland. Even if TVWS networks are not deployed in Finland, the research conducted in the field of TVWS has been and will be beneficial for the development of future 5G wireless networks, as spectrum sharing and incumbent protection will play a huge role in the networks operating on frequencies below 6 GHz. The WISE project was part of the Tekes Trial Environment for Cognitive Radio and Network programme (2011–2015). Currently the Turku test network is being used to study licensed shared use in the UHF band in the Future of UHF project [29], which belongs to the Tekes 5th Gear programme [30] and to the 5G Test Network Finland ecosystem [31].

## ACKNOWLEDGEMENTS

The research has been carried out during the WISE and WISE2 projects funded by Tekes (The Finnish Funding Agency for Innovation).

## REFERENCES

[1] WISE project, WISE project web page, http://wise.turkuamk.fi/. (Accessed 24 March 2016).
[2] CEPT/ECC, SE43 working group, http://www.cept.org/ecc/groups/ecc/closed-groups/se-43. (Accessed 24 March 2016).
[3] CEPT/ECC, SE(12)040 Annex 04: list of WG SE project teams and their terms of reference, http://www.cept.org/Documents/wg-se/4854/SE(12)040-annex-04_Project-Teams-and-ToR. (Accessed 24 March 2016).
[4] ECC Report 159: Technical and operational requirements for the possible operation of cognitive radio systems in the 'white spaces' of the frequency band 470–790 MHz, Jan. 2011.
[5] ECC Report 185: Complementary Report to ECC Report 159. Further definition of technical and operational requirements for the operation of white space devices in the band 470–790 MHz, Jan. 2013.
[6] ECC Report 186: Technical and operational requirements for the operation of white space devices under geo-location approach, Jan. 2013.
[7] Official Journal of the European Union, Directive 1999/5/EC of the European Parliament and of the Council of 9 March 1999 on radio equipment and telecommunications terminal equipment and the mutual recognition of their conformity, http://eur-lex.europa.eu/legal-content/EN/TXT/?uri=CELEX:31999L0005. (Accessed 24 March 2016).
[8] ETSI, White Space Devices (WSD); Wireless Access Systems operating in the 470 MHz to 790 MHz frequency band; Harmonized EN covering the essential requirements of article 3.2 of the R&TTE Directive (EN301598), Version v1.1.1 (Accessed 16 March 2016), Apr. 2014.
[9] Directive 2014/53/EU of the European Parliament and of the Council of 16 April 2014 on the harmonisation of the laws of the Member States relating to the making available on the market of radio equipment and repealing Directive 1999/5/EC Text with

EEA relevance, http://eur-lex.europa.eu/legal-content/EN/TXT/?uri=CELEX:32014L0053. (Accessed 24 March 2016).

[10] Fairspectrum, Press release: Fairspectrum provides TV white space database for Europe's first geolocation radio license, Aug. 2012, http://fairspectrum.com/propagating-thoughts/pressrelease%20fairspectrumprovidestvwhitespacedatabaseforeurope-%E2%80%99%20sfirstgeolocationradiolicense. (Accessed 7 June 2016).

[11] Carlson rural connect, http://www.rsigcg.com/files/RuralConnect.pdf. (Accessed 24 March 2016).

[12] IETF, Protocol to access white-space (PAWS) databases RFC 7545, https://datatracker.ietf.org/doc/rfc7545/. (Accessed 24 March 2016).

[13] Jaakko Ojaniemi, et al., Optimal field measurement design for radio environment mapping, in: 2013 47th Annual Conference on Information Sciences and Systems, CISS 2013, 2013.

[14] Jaakko Ojaniemi, et al., A practical method for combining multivariate data in radio environment mapping, in: IEEE International Symposium on Personal, Indoor and Mobile Radio Communications, PIMRC, 2013, pp. 729–733.

[15] ITU-R recommendation P.1812-4: a path-specific propagation prediction method for point-to-area terrestrial services in the VHF and UHF bands, July 2015, http://www.itu.int/rec/R-REC-P.1812-2-201202-S/. (Accessed 24 March 2016).

[16] CEPT/SE43(11)36 WSD Maximum power considerations 10th SE43 meeting, Bologna, Italy, July 2011, http://www.cept.org/Documents/se-43/596/SE43_11_36_WSD_Max_Power. (Accessed 24 March 2016).

[17] CEPT/ECC SE43(11) info 08 WSD maximum power measurement report, 10th SE43 meeting, Bologna, Italy, July 2011, http://www.cept.org/Documents/se-43/593/SE43_11_Info08_WSD_Max_Power_Measurement_Report. (Accessed 24 March 2016).

[18] CEPT/ECC SE43(11)81 wise-project measurement report: WSD maximum power indoor measurements in Turku test network, 12th SE43 meeting, Cambridge, UK, December 2011, http://www.cept.org/Documents/se-43/3426/SE43%5C%2811%5C%2981AP1_Wise-Project-Measurement-Report-WSD-maximum-Power-Indoor-Measurements-in-Turku-Test. (Accessed 24 March 2016).

[19] CEPT/ECC SE43(11)81 WSD maximum power measurements in indoor 2 m reference geometry, 12th SE43 meeting, Cambridge, UK, December 2011, http://www.cept.org/Documents/se-43/3422/SE43%5C%2811%5C%2981_WSD-maximum-power-measurements-in-indoor-2-m-reference-geometry. (Accessed 24 March 2016).

[20] CEPT/ECC SE43(11)82 PMSE protection measurements in Helsinki City Theatre, 12th SE43 meeting, Cambridge, UK, December 2011, www.cept.org/Documents/se-43/3424/SE43(11)82_PMSE-protection-measurements-in-Helsinki-City-Theatre. (Accessed 28 June 2016).

[21] Wise-project measurement report: PMSE measurements in Helsinki City Theatre, 12th SE43 meeting, Cambridge, UK, December 2011, http://www.cept.org/Documents/se-43/3425/SE43(11)82AP1_Wise-Project-Measurement-Report-PMSE-Measurements-in-Helsinki-City-Theatre. (Accessed 28 June 2016).

[22] ECC report 148, Measurements on the performance of DVB-T receivers in the presence of interference from the mobile service, June 2010.

[23] ITU-R Document 6E/64-E, The ESR5 criterion for the Assessment of DVB-T transmission Quality, Apr. 2004.

[24] Pekka Talmola, et al., Field measurements of WSD-DTT protection ratios over outdoor and indoor reference geometries, in: 2012 7th International Conference on Cognitive Radio Oriented Wireless Networks and Communications, CROWNCOM, 2012.

[25] Recommendation ITU-R BT.1368-12: planning criteria, including protection ratios, for digital terrestrial television services in the VHF/UHF bands, Feb. 2015, http://www.itu.int/rec/R-REC-BT.1368-12-201502-I/en. (Accessed 30 May 2016).

[26] Matti Simolin, Video Surveillance over Cognitive Radio Technology, Bachelor's thesis, Turku University of Applied Sciences, 2013.

[27] Sami Elo, Testing Cognitive Radio System Applications, Bachelor's thesis, Turku University of Applied Sciences, 2015.

[28] Janne Isosaari, Testing Cognitive Radio System, Bachelor's thesis, Turku University of Applied Sciences, 2013.

[29] FUHF project, FUHF project web page, http://fuhf.turkuamk.fi/. (Accessed 24 April 2016).

[30] Tekes' 5th Gear programme, https://www.tekes.fi/en/programmes-and-services/tekes-programmes/5thgear/projects/. (Accessed 24 March 2016).

[31] 5G test network Finland, http://5gtnf.fi/. (Accessed 30 June 2016).

Chapter 8

# Techno-Economic Evaluation for TV White Spaces

**Carlos F.M. e Silva and F. Rodrigo P. Cavalcanti**
*Wireless Telecom Research Group (GTEL), Federal University of Ceará (UFC), Fortaleza, CE, Brazil*

## 8.1 INTRODUCTION

Over the last few years the tremendous success of smartphones and tablet computers along with the massive use of rich data applications such as video streaming, social networking and online gaming (to name just a few) are stressing the capacity of wireless systems. On the one hand, users keep demanding more and more mobile broadband data, generating a growth of around 50% each year [1]. On the other hand, by the year 2018 Internet of Things (IoT) devices (consumer electronics, connected machines, sensors, utility meters, etc.) are expected to surpass 'traditional' smartphones and tablets driven by new use cases [2].

Despite the worldwide deployment of Fourth Generation (4G)/Long Term Evolution (LTE) cellular and Wireless Fidelity (Wi-Fi) private networks that are making a significant effort to keep up with these demands, the expectation is that they will fall short of the required capacity. Hence, the first thoughts about future Fifth Generation (5G) systems are now arising [3–6].

There is a common belief that these systems will be essentially heterogeneous, following a multi-tier architecture [7], and the move to very high central carrier frequencies (30 to 300 GHz) seems to be inevitable due to the lack of available spectrum at lower frequencies, namely in Ultra High Frequency (UHF)/microwave bands. As such, current research topics include, e.g. massive Multiple Input, Multiple Output (MIMO) [8], millimetre wave bands [9], Machine Type Communications (MTCs) [10,11], Heterogeneous Networks (HetNets) [12] and spectrum sharing [13]. Nonetheless, 5G systems are only expected to be standardised by the year 2020.

Moreover, it is well-known that the electromagnetic spectrum is a scarce resource, being already occupied or reserved for certain services/applications. However, according to measurement campaigns [14–16] in the range of 30 MHz

**TV White Space Communications and Networks. http://dx.doi.org/10.1016/B978-0-08-100611-5.00011-4**

to 3 GHz, the maximum spectrum utilisation is in general around 25%, and the average use stays well below that fraction.

Traditional spectrum markets and static spectrum management originated this paradoxal situation: the spectrum is occupied without being used! Long-term licences (15 years or more)—intended to prevent harmful interference between various radio licensees—means that the spectrum is underutilised in many geographical areas. Thus, traditional markets cannot guarantee the efficient and flexible allocation of spectrum.

The recent global move from analogue to digital Television (TV) opened a wide range of new spectrum management opportunities. Since digital TV is more spectrally efficient, the TV band (470 to 862 MHz) has been cleared (790 to 862 MHz) for cellular use and refurbished (470 to 790 MHz) in slots of 6 or 8 MHz (USA or Europe, respectively) TV channels; this is called the digital dividend [18]. (The actual range of frequencies within the TV band depends on the country. For example, in the United Kingdom (UK) the range is from 470 to 862 MHz [17], while in the United States of America (USA) it ranges from 475 to 806 MHz [14]. In the same manner, in the UK the band from 790 to 862 MHz has been cleared for cellular use, while in USA the cleared band is from 698 to 806 MHz.)

Due to the need for managing interference in the TV broadcast network from neighbouring towers, channels are interleaved both in frequency and space. The vacant/unused channels are known as TV White Spaces (TVWS). In [19] the European Conference of Postal and Telecommunications Administrations (CEPT) identifies white spaces as ‘a part of the spectrum, which is available for a radiocommunication application (service, system) at a given time in a given geographical area on a non-interfering/non-protected basis with regard to other services with a higher priority on a national basis’.

More specifically, the interleaved spectrum or TVWS arises because in a multiple frequency network any TV channel is carried on a number of different frequency channels around the service zone. On any given frequency channel there will be a geographical area where the use of high-power broadcasting is not possible because of the interference it would cause, but low/moderate power applications may operate, provided they are carefully designed to be compatible with the primary TV band users and other secondary users (e.g. Programme Making and Special Events (PMSE) or wireless microphones) [20]. (Incumbents are primary spectrum users since they own the spectrum, due to regulatory decisions or traditional (primary) spectrum market, which follows the long-term licensing schema. Secondary users are the spectrum users who may only use it when no primary users are using it, employing opportunistic methods.)

The TV portion of spectrum is famous for its advantageous properties: signal waves travel further and penetrate in buildings easily, which are qualities

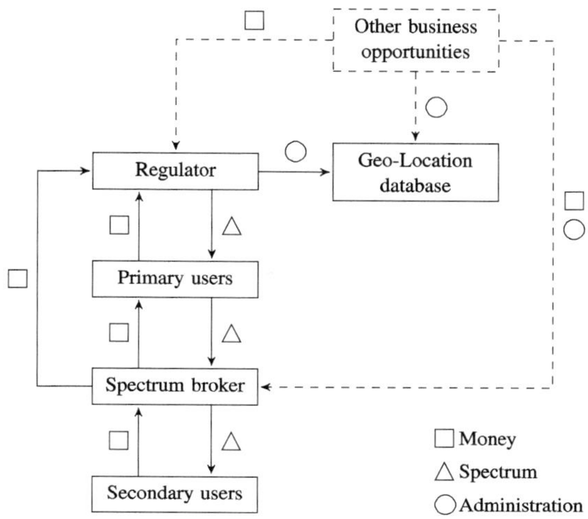


**FIGURE 8.1** Flows for the secondary spectrum market; along with spectrum, money, rights and obligations are also exchanged.

already exploited by TV broadcasters. As such, the implementation of a flexible secondary spectrum market for the use of TVWS becomes desirable. However, primary (licensed or incumbent) users are reluctant because they fear strong and unpredictable interference from secondary systems.

There are, instead, possible sensing capabilities from White Space Devices (WSDs) (or Cognitive Radio (CR) devices that operate in TVWS. These devices are empowered by reconfigurable capabilities that enable them to obtain the technical, regulatory, and operational requirements, so that they can adapt themselves to the surrounding environment and are able to use the spectrum [21].); a secondary spectrum market where channels, spectrum rights, and obligations are traded in a real-time fashion, as clearly stated by the National Regulator Authority (NRA); geolocation databases [22,23] and spectrum brokers [20,24] which should limit that fear of interference. For example, Fig. 8.1 shows the possible flows of money, spectrum allocations, and administrative information for the secondary spectrum market [25].

In this work we describe the methodology and perform a techno-economic evaluation of three different deployment scenarios (assuming traditional and secondary markets) where LTE legacy and TVWS frequency carriers are available. The scenarios are the following:

- Deployment scenario 1: traditional spectrum market for deployment using only the legacy carrier of 2.6 GHz;
- Deployment scenario 2: traditional spectrum market for deployment using only the TVWS carrier of 700 MHz;

- Deployment scenario 3: flexible spectrum market for deployment using the legacy carrier plus an additional TVWS carrier.

The frequency carriers listed above were chosen because 2.6 GHz is one of the most common frequencies used in LTE deployments around the world and although it is derived from the digital dividend, the 700 MHz frequency is nowadays being traded in some countries for cellular use.

The remaining sections are organised as follows: in Section 8.2 we present the methodology followed and the metric used, in Section 8.3 we show the calculations required to determine the number of network elements, in Section 8.4 we estimate the cost of each element considered, in Section 8.5 we present the main results and discuss their implications, and finally Section 8.6 we present our conclusion and discuss future possible work.

## 8.2 METHOD FOR TECHNO-ECONOMIC EVALUATION

The economic metric used for techno-economic evaluation is the Net Present Value (NPV) since it is one of the most widely used and reliable indicators to evaluate the feasibility of a project. Hence, in order to compute the NPV we must clearly understand what the revenues and costs are in the form of a cash flow (CF). The deployment scenarios considered in this work are for a single operator (there is no renting income). In this case, the only revenues are the sum of all payments made by clients for the services they have contracted.

Generally, the costs of building and running a broadband network can be divided into Capital Expenditure (CAPEX) and Operational Expenditure (OPEX). CAPEX includes the investments in the network infrastructure and devices, as well as the hardware required for Operational and Maintenance (O&M) functions, such as network management and billing/charging systems. OPEX includes the labour costs and expenses originated from running and managing the network, as well as costs related to marketing, sales, and client care.

Consider for example that we want to buy a new car. CAPEX would be the money paid to the seller (either paid all at once or by paying a fixed amount per month) plus the money to buy a garage to park the car. In addition to that, OPEX can be considered as the money for renting the garage, paying for the general maintenance and car repair, plus the money spent for replacing the wheels and filling the car tank with fuel.

The OPEX related with a certain project is often more difficult to predict than the CAPEX. This is especially true when new technologies are considered, as the previous experience and/or data are not available.

When revenues, investments and all operational costs are estimated for each year of the period under study ($N$ years), the cash flow $CF(n)$, $n \in \{0, 1, \ldots, N\}$, can be established as

$$\mathrm{CF}(n) = \mathrm{Revenues}(n) - \mathrm{CAPEX}(n) - \mathrm{OPEX}(n). \tag{8.1}$$

The time-value of money and risks are considered in the discount rate $r$ (sometimes also referred to as interest rate). Therefore, the discounted cash flow is defined as

$$\mathrm{DCF}(n) = \frac{\mathrm{CF}(n)}{(1+r)^n}. \tag{8.2}$$

Finally, the sum of all discounted cash flows is known as the NPV,

$$\mathrm{NPV} = \mathrm{CF}(0) + \sum_{n=1}^{N} \mathrm{DCF}(n) = \sum_{n=0}^{N} \mathrm{DCF}(n). \tag{8.3}$$

The NPV (see [26] for more information about the historical use of NPV) is a metric used to measure the value of a project: if NPV is positive the investment adds value to the company and the project is profitable; instead, if NPV is negative the investment subtracts value from the company and the project will be abandoned or reformulated; finally, if NPV is equal to zero then the investment neither gains nor loses value for the company, thus the final decision will be based on other criteria, e.g. strategic position. In [27] the decision criteria is further explained based on the analysis of project risks.

Hence, the NPV rule states that a company should invest in any project with a positive NPV, and since it incorporates the discount rate, it represents the expected return that is forgone by investing in the project rather than in comparable financial securities. Other methods, such as the Internal Rate of Return (IRR) rule or the payback rule also used in profitability assessment, have some pitfalls and deficiencies [28] when compared with the NPV method, and therefore are not used in this work.

Additionally, the first year or year of implementation (y0) mainly deals with investments. As such, the CF(0) is expected to be negative; for educational purposes, CF(0) is commonly placed to the left of the sum to emphasise its role as (minus) the investment.

### 8.2.1 Scenario Description

For a comparative study on the expenditures of deploying and operating an LTE network over TVWS against legacy carrier frequencies, knowledge about the scenario where the LTE network will be deployed is required. In this context, we present the characterisation of the area under study in terms of occupied area, population density, client base and number of active users. Moreover,

**FIGURE 8.2** Munich area under consideration.

we consider an LTE network with its main entities (access, transport and core networks), and a spectrum broker and geolocation database for the secondary spectrum market of TVWS.

### 8.2.1.1 Geographical Area and Population

The area for evaluation is a square of 50 × 50 km around the city of Munich in Germany, which includes its metropolitan area. The red square in Fig. 8.2 shows the different zones that were considered: Munich (centre), Ebersberg, Erding, Munich (district), Freising and Dachau. Munich was chosen because it characterises a typical European city, but this study could be performed for any other city.

In Table 8.1 the most relevant information about each zone for network planning is presented, namely, occupied area, population density and number of active users. The population density is very important for classifying areas as urban, suburban and rural. Therefore, it provides the basis for projecting the likely number of users in different areas, which is shown in the table as the percentage of clients and active users according to the estimates made in [29]. (Note that in the calculations regarding the number of clients, the migratory movement of people, which is difficult to predict, is not considered.) These parameters are used in turn as input to determine the required network elements to meet coverage, capacity and quality objectives for the different offered services.

**TABLE 8.1** Characterisation of the Area Under Consideration. In any Region, the Percentage of Clients Is Assumed to Be 18%[b]

| Region | Borough | Area ($km^2$)[a] | Population[a] | Density (population/$km^2$) | Act. users (%)[c] | Num. of users |
|---|---|---|---|---|---|---|
| Munich (center) | Altstadt-Lehelv (1) | 3.16 | 18876 | 5973 | 5 | 170 |
| | Ludwigsvorstadt-Isarvorstadt (2) | 4.39 | 45736 | 10418 | 5 | 412 |
| | Maxvorstadt (3) | 4.29 | 46058 | 10736 | 5 | 415 |
| | Schwabing-West (4) | 4.37 | 59553 | 13628 | 5 | 536 |
| | Au-Haidhausen (5) | 4.22 | 54382 | 12887 | 5 | 489 |
| | Sendling (6) | 3.94 | 37146 | 9428 | 5 | 334 |
| | Sendling-Westpark (7) | 7.81 | 50903 | 6518 | 5 | 458 |
| | Schwanthalerhöhe (8) | 2.07 | 26103 | 12610 | 5 | 235 |
| | Neuhausen-Nymphenburg (9) | 12.92 | 84604 | 6548 | 5 | 761 |
| | Moosach (10) | 11.09 | 47754 | 4306 | 5 | 430 |
| | Milbertshofen-Am Hart (11) | 13.37 | 66992 | 5011 | 5 | 603 |
| | Schwabing-Freimann (12) | 25.67 | 62430 | 2432 | 5 | 562 |
| | Bogenhausen (13) | 23.71 | 75657 | 3191 | 5 | 681 |
| | Berg am Laim (14) | 6.31 | 39009 | 6182 | 5 | 351 |
| | Trudering-Riem (15) | 22.45 | 53915 | 2402 | 5 | 485 |
| | Ramersdorf-Perlach (16) | 19.90 | 102689 | 5160 | 5 | 924 |
| | Obergiesing (17) | 5.71 | 47007 | 8232 | 5 | 423 |
| | Untergiesing-Harlaching (18) | 8.06 | 48075 | 5965 | 5 | 433 |
| | Thalkirchen-Obersendling-Forstenried-Fürstenried-Solln (19) | 17.75 | 80701 | 4547 | 5 | 726 |
| | Hadern (20) | 9.23 | 44993 | 4875 | 5 | 405 |
| | Pasing-Obermenzing (21) | 16.50 | 63763 | 3864 | 5 | 574 |
| | Aubing-Lochhausen-Langwied (22) | 34.06 | 37857 | 1111 | 5 | 341 |
| | Allach-Untermenzing (23) | 15.45 | 27730 | 1795 | 5 | 250 |

*continued on next page*

**TABLE 8.1** (*continued*)

| Region | Borough | Area ($km^2$)[a] | Population[a] | Density (population/$km^2$) | Act. users (%)[c] | Num. of users |
|---|---|---|---|---|---|---|
| | Feldmoching-Hasenbergl (24) | 28.71 | 54245 | 1889 | 5 | 488 |
| | Laim (25) | 5.29 | 50082 | 9467 | 5 | 451 |
| Subtotal | | 310.43 | 1326260 | 159175 | | 11936 |
| Ebersberg | Ebersberg | 40.84 | 11394 | 279 | 3 | 62 |
| | Grafing | 29.57 | 12865 | 435 | 3 | 69 |
| | Anzing | 16.19 | 3583 | 221 | 3 | 19 |
| | Aßling | 31.38 | 4299 | 137 | 1 | 15 |
| | Baiern | 19.96 | 1513 | 76 | 1 | 5 |
| | Bruck | 21.6 | 1167 | 54 | 1 | 4 |
| | Egmating | 19.16 | 2142 | 112 | 1 | 8 |
| | Emmering | 17.22 | 1440 | 84 | 1 | 5 |
| | Forstinning | 12.26 | 3505 | 286 | 3 | 19 |
| | Frauenneuharting | 22.68 | 1475 | 65 | 1 | 5 |
| | Glonn | 30.24 | 4439 | 147 | 1 | 16 |
| | Hohenlinden | 17.32 | 2822 | 163 | 1 | 10 |
| | Kirchseeon | 17.91 | 9593 | 536 | 3 | 52 |
| | Markt Schwaben | 10.87 | 11911 | 1096 | 5 | 107 |
| | Moosach | 18.21 | 1461 | 80 | 1 | 5 |
| | Oberpframmern | 18.47 | 2188 | 118 | 1 | 8 |
| | Pliening | 22.8 | 5159 | 226 | 3 | 28 |
| | Poing | 12.89 | 13425 | 1042 | 5 | 121 |
| | Steinhöring | 36.29 | 3912 | 108 | 1 | 14 |
| | Vaterstetten | 34.18 | 22070 | 646 | 3 | 119 |
| | Zorneding | 23.77 | 8836 | 372 | 3 | 48 |
| Subtotal | | 473.81 | 129199 | 6281 | | 741 |

*continued on next page*

**TABLE 8.1** *(continued)*

| Region | Borough | Area ($km^2$)[a] | Population[a] | Density (population/$km^2$) | Act. users (%)[c] | Num. of users |
|---|---|---|---|---|---|---|
| Erding | Dorfen | 99.6 | 13723 | 138 | 1 | 49 |
| | Erding | 54.64 | 34514 | 632 | 3 | 186 |
| | Berglern | 19.89 | 2568 | 129 | 1 | 9 |
| | Bockhorn | 47.15 | 3542 | 75 | 1 | 13 |
| | Buch am Buchrain | 22.75 | 1412 | 62 | 1 | 5 |
| | Eitting | 35.63 | 2415 | 68 | 1 | 9 |
| | Finsing | 23.17 | 4319 | 186 | 1 | 16 |
| | Forstern | 15.38 | 3254 | 212 | 3 | 18 |
| | Fraunberg | 42.37 | 3354 | 79 | 1 | 12 |
| | Hohenpolding | 27.42 | 1476 | 54 | 1 | 5 |
| | Inning am Holz | 11.83 | 1439 | 122 | 1 | 5 |
| | Isen | 43.78 | 5293 | 121 | 1 | 19 |
| | Kirchberg | 17.09 | 920 | 54 | 1 | 3 |
| | Langenpreising | 27.49 | 2640 | 96 | 1 | 10 |
| | Lengdorf | 33.9 | 2743 | 81 | 1 | 10 |
| | Moosinning | 39.96 | 5521 | 138 | 1 | 20 |
| | Neuching | 19.66 | 2411 | 123 | 1 | 9 |
| | Oberding | 64.7 | 5384 | 83 | 1 | 19 |
| | Ottenhofen | 10.27 | 1843 | 179 | 1 | 7 |
| | Pastetten | 22.05 | 2478 | 112 | 1 | 9 |
| | Sankt Wolfgang | 46.32 | 4312 | 93 | 1 | 16 |
| | Steinkirchen | 18.08 | 1153 | 64 | 1 | 4 |
| | Taufkirchen (Vils) | 70.18 | 8980 | 128 | 1 | 32 |
| | Walpertskirchen | 18.45 | 2067 | 112 | 1 | 7 |

*continued on next page*

**TABLE 8.1** (*continued*)

| Region | Borough | Area ($km^2$)[a] | Population[a] | Density (population/$km^2$) | Act. users (%)[c] | Num. of users |
|---|---|---|---|---|---|---|
| | Wartenberg | 17.88 | 4783 | 268 | 3 | 26 |
| | Wörth | 21.05 | 4467 | 212 | 3 | 24 |
| Subtotal | | 870.69 | 127011 | 3620 | | 542 |
| Munich (district) | Garching | 28.16 | 15860 | 563 | 3 | 86 |
| | Unterschleißheim | 14.93 | 26416 | 1769 | 5 | 238 |
| | Aschheim | 28.04 | 7665 | 273 | 3 | 41 |
| | Aying | 44.98 | 4499 | 100 | 1 | 16 |
| | Baierbrunn | 7.21 | 2944 | 408 | 3 | 16 |
| | Brunnthal | 26.92 | 4765 | 177 | 1 | 17 |
| | Feldkirchen | 6.41 | 6389 | 997 | 3 | 35 |
| | Gräfelfing | 9.58 | 12870 | 1343 | 5 | 116 |
| | Grasbrunn | 23.59 | 6411 | 272 | 3 | 35 |
| | Grünwald | 7.63 | 11057 | 1449 | 5 | 100 |
| | Haar | 12.9 | 19430 | 1506 | 5 | 175 |
| | Hohenbrunn | 16.82 | 8954 | 532 | 3 | 48 |
| | Höhenkirchen-Siegertsbrunn | 15.19 | 9803 | 645 | 3 | 53 |
| | Ismaning | 40.19 | 15389 | 383 | 3 | 83 |
| | Kirchheim | 15.51 | 12463 | 804 | 3 | 67 |
| | Neubiberg | 5.77 | 13938 | 2416 | 5 | 125 |
| | Neuried | 9.63 | 8411 | 873 | 3 | 45 |
| | Oberhaching | 26.6 | 12939 | 486 | 3 | 70 |
| | Oberschleißheim | 30.6 | 11296 | 369 | 3 | 61 |
| | Ottobrunn | 5.53 | 20105 | 3636 | 5 | 181 |
| | Planegg | 10.68 | 10415 | 975 | 3 | 56 |
| | Pullach | 7.41 | 8733 | 1179 | 5 | 79 |

*continued on next page*

**TABLE 8.1** *(continued)*

| Region | Borough | Area ($km^2$)[a] | Population[a] | Density (population/$km^2$) | Act. users (%)[c] | Num. of users |
|---|---|---|---|---|---|---|
| | Putzbrunn | 11.17 | 6035 | 540 | 3 | 33 |
| | Sauerlach | 48.49 | 7128 | 147 | 1 | 26 |
| | Schäftlarn | 16.71 | 5564 | 333 | 3 | 30 |
| | Straßlach-Dingharting | 28.34 | 2963 | 105 | 1 | 11 |
| | Taufkirchen | 22.02 | 17868 | 811 | 3 | 96 |
| | Unterföhring | 12.8 | 9931 | 776 | 3 | 54 |
| | Unterhaching | 8.73 | 22774 | 2609 | 5 | 205 |
| Subtotal | | 542.54 | 323015 | 26477 | | 2197 |
| Freising | Freising | 88.45 | 45223 | 511 | 3 | 244 |
| | Moosburg | 43.86 | 17554 | 400 | 3 | 95 |
| | Allershausen | 26.56 | 5015 | 189 | 1 | 18 |
| | Attenkirchen | 16.16 | 2674 | 165 | 1 | 10 |
| | Au in der Hallertau | 54.99 | 5615 | 102 | 1 | 20 |
| | Eching | 37.83 | 13275 | 351 | 3 | 72 |
| | Fahrenzhausen | 37.64 | 4624 | 123 | 1 | 17 |
| | Gammelsdorf | 21.63 | 1578 | 73 | 1 | 6 |
| | Haag an der Amper | 21.7 | 2869 | 132 | 1 | 10 |
| | Hallbergmoos | 35.06 | 9266 | 264 | 3 | 50 |
| | Hohenkammer | 25.74 | 2299 | 89 | 1 | 8 |
| | Hörgertshausen | 21.47 | 1904 | 89 | 1 | 7 |
| | Kirchdorf an der Amper | 32.99 | 2752 | 83 | 1 | 10 |
| | Kranzberg | 39.5 | 4011 | 102 | 1 | 14 |
| | Langenbach | 26.89 | 3910 | 145 | 1 | 14 |
| | Marzling | 20.5 | 3099 | 151 | 1 | 11 |
| | Mauern | 24.14 | 2780 | 115 | 1 | 10 |

*continued on next page*

**TABLE 8.1** (*continued*)

| Region | Borough | Area ($km^2$)[a] | Population[a] | Density (population/$km^2$) | Act. users (%)[c] | Num. of users |
|---|---|---|---|---|---|---|
| | Nandlstadt | 34.31 | 4977 | 145 | 1 | 18 |
| | Neufahrn bei Freising | 45.51 | 19046 | 419 | 3 | 103 |
| | Paunzhausen | 12.73 | 1555 | 122 | 1 | 6 |
| | Rudelzhausen | 40.82 | 3202 | 78 | 1 | 12 |
| | Wang | 31.19 | 2462 | 79 | 1 | 9 |
| | Wolfersdorf | 26.01 | 2414 | 93 | 1 | 9 |
| | Zolling | 34.59 | 4271 | 123 | 1 | 15 |
| Subtotal | | 800.27 | 166375 | 4145 | | 787 |
| Dachau | Dachau | 34.85 | 42506 | 1220 | 5 | 383 |
| | Altomünster | 75.79 | 7405 | 98 | 1 | 27 |
| | Bergkirchen | 59.99 | 7195 | 120 | 1 | 26 |
| | Erdweg | 36.03 | 5595 | 155 | 1 | 20 |
| | Haimhausen | 26.73 | 4893 | 183 | 1 | 18 |
| | Hebertshausen | 29.6 | 5321 | 180 | 1 | 19 |
| | Hilgertshausen-Tandern | 28.64 | 3158 | 110 | 1 | 11 |
| | Karlsfeld | 15.55 | 18293 | 1176 | 5 | 165 |
| | Markt Indersdorf | 68.6 | 9377 | 137 | 1 | 34 |
| | Odelzhausen | 30.48 | 4312 | 141 | 1 | 16 |
| | Petershausen | 32.82 | 6110 | 186 | 1 | 22 |
| | Pfaffenhofen an der Glonn | 20.9 | 1805 | 86 | 1 | 6 |
| | Röhrmoos | 31.74 | 6281 | 198 | 1 | 23 |
| | Schwabhausen | 30.23 | 6199 | 205 | 3 | 33 |

*continued on next page*

**TABLE 8.1** (*continued*)

| Region | Borough | Area ($km^2$)[a] | Population[a] | Density (population/$km^2$) | Act. users (%)[c] | Num. of users |
|---|---|---|---|---|---|---|
| | Sulzemoos | 19.04 | 2668 | 140 | 1 | 10 |
| | Vierkirchen | 19.39 | 4278 | 221 | 3 | 23 |
| | Weichs | 18.72 | 3151 | 168 | 1 | 11 |
| Subtotal | | 579.1 | 138547 | 4725 | | 846 |
| Total | | 3576.84 | 2210407 | 204424 | | 17048 |

[a] *The geographical area and population information is as in the latest demographical report (census) available for the year 2012.*

[b] *The number of clients for a given operator (market sharing) is intrinsically related with its market position against competitors for the area under consideration. This value is for the first year of deployment.*

[c] *The number of active users is given accordingly with the geographical classification. As such, for urban, sub-urban, and rural areas, respectively, 5, 3, and 1% of the clients are active [29], which is used as input for coverage and capacity planning.*

From the geographical and population input information, the population density and the geographical classification are defined, which is needed in order to start the coverage (e.g. selection of propagation models) and capacity (e.g. number of active users) dimensioning exercises. More specifically, the area is considered rural if the population density is lower than 200 people/km$^2$, suburban when population density ranges from 200 to 1000 people/km$^2$ and urban if the population density is above 1000 people/km$^2$.

The classification presented here follows a simple rule; however, giving an absolutely objective criteria to the classification of a geographical area into urban, suburban and rural zones is probably impossible because any method requires a choice of thresholds, which is subjective to a certain extent. However, the combination of other criteria can be used instead [30,31]. Fig. 8.3 depicts the Munich area (red square in Fig. 8.2) and the geographical classification for each zone.

### *8.2.1.2 System Overview*

In Fig. 8.4 the LTE network with additional TVWS entities is shown. Here, we skip a detailed explanation of these elements and focus on how they are integrated in our evaluation. A detailed description about LTE systems and the functionalities of each of its network entities can be found in [32]. In the same manner, details regarding the spectrum broker and geolocation database are presented in [24]. However, it is important to highlight that the implementation of a flexible spectrum market is based on the premise that regulators adopt unambiguous rules which set out clearly the rights and obligations of incumbents and any secondary system that use TVWS on a licensed basis. Also, the use of a spectrum broker limits the possible generated interference in both primary and secondary systems.

Like any cellular network, LTE systems can be divided into access network, transport network and core network. The access network, or Evolved Universal Terrestrial Radio Access Network (E-UTRAN), is constituted of Base Stations (BSs) or Evolved Node Bs (eNBs) (using LTE terminology) interconnected through the X2 interface, and is the link between clients and other network entities. The transport network is the connection element through the S1 interface between the access and core networks. It delivers the data or control packets, and for this work we considered it to be based on optical fibre. (The use of optical fibre is reasonable for densely populated cities. However, in rural areas the use of microwave links may be seen as an alternative choice.) The core network, or Evolved Packet Core (EPC), is the central part of the whole network and provides various services for clients who are connected by the access network, such as packet routing, mobility management, authentication, accounting, and access

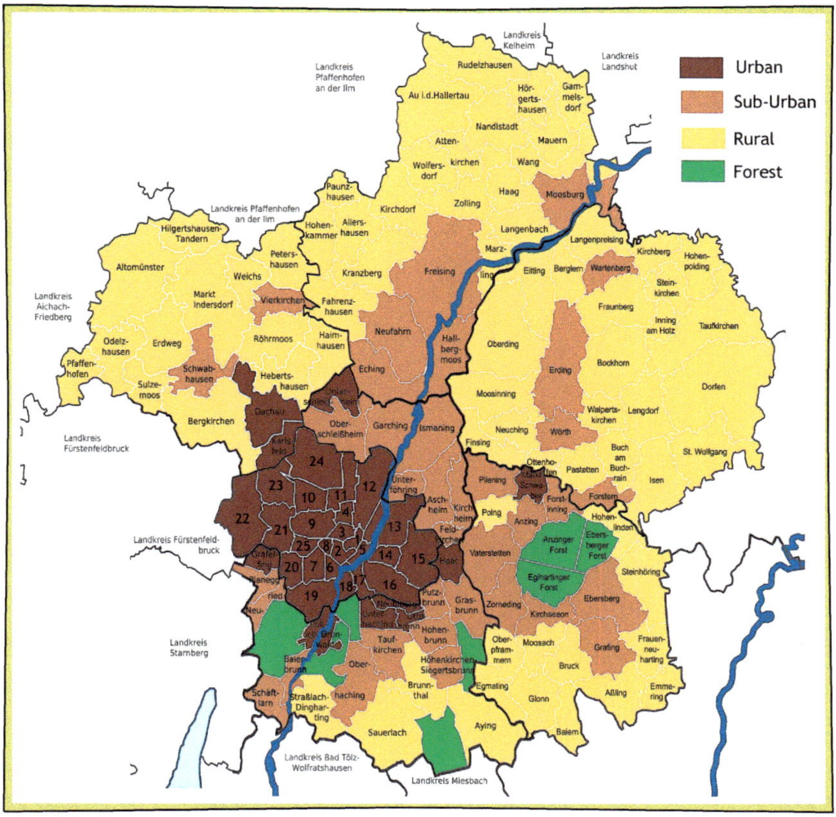


FIGURE 8.3 Munich area with the geographical classification per zone.

to external networks. The EPC is, in a general sense, constituted by the Mobility Management Entity (MME), Policy and Charging Rules Function (PCRF), Service Gateway (S-GW), and Packet Gateway (P-GW). Other network entities, such as a data centre with various types of databases where the profiles of clients are stored, and the Operational and Maintenance (O&M) centre, also make up the LTE network.

The EPC may be limited by two factors: first, the number of BSs that can be connected to a single MME; second, the amount of traffic that can be processed at the same time by the S-GW and P-GW. It is assumed that if a new S-GW–P-GW pair is required, then a new EPC is required. However, in many cases, a new MME is not required because it only processes control information and can handle a large number of connected BSs.

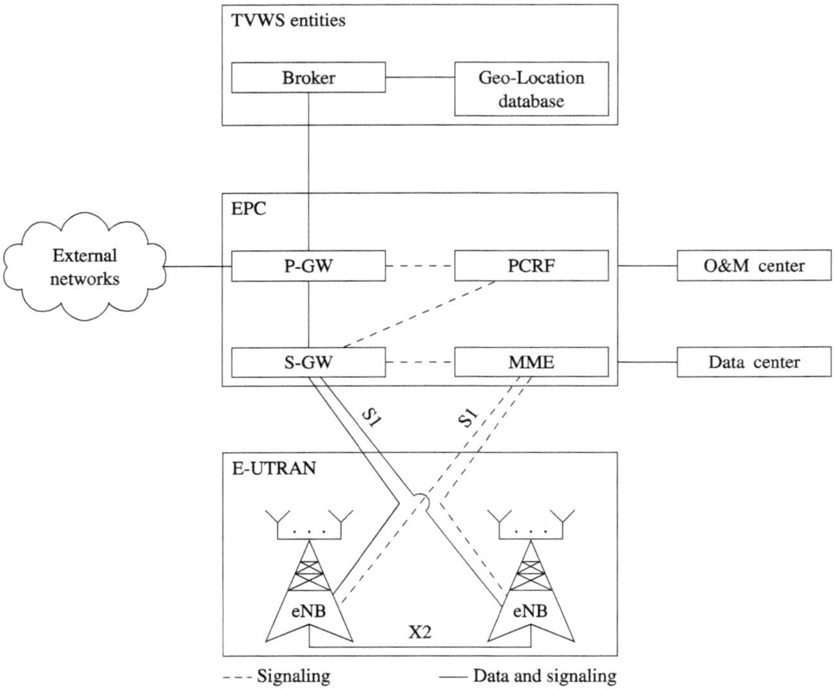


**FIGURE 8.4** LTE network with additional TVWS entities for a flexible spectrum market.

Moreover, the BSs are connected to each other through the X2 interface, which mostly carries control information, and can be seen as a logical link. Practically speaking, and assuming that the delay in the transport network can be neglected, the X2 interface can be implemented through the core network and multiplexed in only one optical fiber. Therefore, from each BS to the core network we only assume one optical fibre link, which has enough capacity to deliver all the traffic that comes(goes) from(to) the BS to(from) the core network.

Finally, the costs associated with the broker and geolocation database (either investments or running costs) are not considered. However, if they were considered and since the number of these elements does not scale with the network size, their impact would be the same in all three of the network deployments, and for comparative studies the relative gains or losses would not change.

## 8.3 NETWORK INFRASTRUCTURE

The network planning is performed to determine the number of BSs that are needed (in terms of both coverage and capacity) to serve a certain region (i.e. to satisfy users' throughput demands). The number of BSs has a direct influence on the costs for operators because, apart from the direct costs associated with BSs (equipment, antennas, renting sites, cooling systems, etc.), it is also used to calculate the length of optical fibre infrastructure to connect the access and core networks. Moreover, the EPC is also limited by the number of connected BSs.

Here we are also interested in reducing the implementation and operational costs while maintaining the network performance, which can be achieved with the introduction of TVWS, because it is expected that fewer BSs (thus lower costs) are required for the same area or that a certain increase in network costs means more capacity can be delivered.

Therefore, a radio network dimensioning tool [33] for the area under consideration was employed. In [29] the planning process in terms of coverage and capacity is described: a flat network configuration, the percentage of clients and active users per geographical area, throughput per user, cell configuration, block rate, and path loss models are the parameters taken into consideration; for consistency, further details are omitted here, but can be easily found in the referred literature. Hence, the radio network dimensioning aims basically to estimate the number of required BSs.

The following network planning activities are explained in the next paragraphs:

- Radio link budget;
- Coverage and capacity analysis;
- Required number of BSs.

### 8.3.1 Radio Link Budget

The radio link budget is an accounting of all the gains and losses in a transmission system. It looks at the elements that will determine the signal strength arriving at the receiver. The radio link budget includes the following items:

- Transmitter power;
- Antenna gains (receiver and transmitter);
- Antenna feeder losses (receiver and transmitter);
- Path loss;
- Required Signal-to-Interference-plus-Noise Ratio (SINR).

With these parameters the maximum allowable path loss for the required SINR is calculated, which can generally be written as

$$L = \mathrm{Tx}_{\mathrm{Power}} + \mathrm{Tx}_{\mathrm{Gain}} - \mathrm{Tx}_{\mathrm{Loss}} + \mathrm{Rx}_{\mathrm{Gain}} - \mathrm{Rx}_{\mathrm{Loss}} - \mathrm{SINR}_{\mathrm{Required}} - \mathrm{Rx}_{\mathrm{Noise}}. \tag{8.4}$$

Where (see Table 8.2 for the reference values):

**TABLE 8.2 Reference Values for the Radio Link Budget Calculation**

| Parameter | DL (BS-user direction) | UL (user-BS direction) |
|---|---|---|
| $Tx_{Power}$ | 38 dBm | 23 dBm |
| $Tx_{Gain}$ | 18 dBi | 0 dBi |
| $Tx_{Loss}$ | 2 dB | 0 dB |
| $Rx_{Gain}$ | 0 dBi | 18 dBi |
| $Rx_{Loss}$ | 0 dB | 0 dB |
| $SINR_{Required}$ | 1 dB | −2.75 dB |
| $Rx_{Noise\ figure}$ | 7 dB | 2 dB |
| Thermal noise | −107.45 dBm | −107.45 dBm |
| $Rx_{Noise}$ [a] | −100.45 dBm | −105.45 dBm |
| Shadowing margin | 4 dB | 1 dB |

[a] *The receiver noise figure plus thermal noise for a bandwidth of 5 MHz.*

- $L$ is the total path loss between the transmitter signal and the receiver (in dB);
- $Tx_{Power}$ is the power transmitted by the transmitter antenna (in dBm);
- $Tx_{Gain}$ is the gain of transmitter antenna related Equivalent Isotropically Radiated Power (EIRP) (in dBi);
- $Tx_{Loss}$ is the transmitter losses (in dB);
- $Rx_{Gain}$ is the gain of receiver antenna related EIRP (in dBi);
- $Rx_{Loss}$ is the receiver losses (in dB);
- $SINR_{Required}$ is the minimum required SINR for the signal to be received at the receiver with the necessary quality/strength, which in LTE is closely related to the Modulation and Coding Scheme (MCS) and service quality provisioning (in dB);
- $Rx_{Noise}$ is the noise power at receiver antenna (in dBm).

## 8.3.2 Coverage and Capacity Analysis

As explained in [29], coverage planning estimates the resources needed to provide service in the deployment area with the given system parameters without any concern regarding capacity. Therefore, it gives an assessment of the resources needed to cover the area under consideration so that the transmitters and receivers can communicate with each other. The coverage analysis fundamentally remains the most critical step in the design of any network.

An appropriate selection of propagation models is required for the coverage analysis and to estimate the number of BSs needed to fulfil the service requirements in the deployment area. The cell radius of a particular LTE sector is calculated through the maximum allowable losses (maximum allowable path loss using (8.4) minus the shadowing margin) both in Downlink (DL) (BS-user direction) and Uplink (UL) (user-BS direction) for the required SINR. Table 8.2 lists the reference values, based on [34]. The maximum coverage range of a cell is thus the minimum between DL and UL calculations, which are obtained through the Erceg Extended model [35,36] for deployment in 2.6 GHz and the Okumura Hata model [37] for deployment in 700 MHz. (Usually the maximum allowable path loss is limited by the UL because in the user-BS direction the transmission power comes from the terminal, which is significantly lower than that from the BS.)

The Erceg Extended model distinguishes three terrain categories, called A, B and C. Category A is for terrain with the highest path loss (hilly terrain with moderate to heavy tree densities) and is used for urban areas. Category B is characterised by flat terrains with intermediate path loss conditions or suburban areas. Finally, category C is for rural areas where path loss is the lowest (mostly flat terrain with light tree densities). Moreover, with appropriate frequency correction factors the model is 'safely' used in the frequency range from 1 to 4 GHz. Instead, the Okumura Hata model, while comprising the same morphologies as the previous model, is used in the frequency range from 150 to 1500 MHz. Thus, the propagation models are aligned with the network deployment and geographical classification considered for this work.

The required SINR also considers the kind of services provided. For that we project that all active users are served with a Constant Bit Rate (CBR) service of 1 Mbps in DL and 256 kbps in UL. After determining the number of BSs needed to cover the area under consideration, the next step is to analyse the capacity issue.

Each BS is limited in its capacity, i.e. there is a maximum amount of traffic (throughput) that it can handle. In some wireless cellular systems, like LTE, coverage and capacity are interrelated. In such cases, the main indicator of capacity is the SINR distribution over the cell, which is obtained by performing physical level simulations in an LTE simulator [29,38,39]. The SINR distribution is directly mapped into system capacity, which is affected by several factors, e.g. scheduler implementation, MCSs, antenna configuration/diversity and interference levels.

As such, in order to map the inter-cell interference levels, different utilisation factors per geographical area (0.85 for urban, 0.88 for suburban and 0.9 for rural) were used and the maximum throughput of a BS was calculated for the Single Input Single Output (SISO) antenna configuration, the system bandwidth of 5 MHz [40], considering the highest LTE MCS [41] and that each BS has three sectors. (The inter-cell interference is predictably higher in urban areas than in suburban or rural areas.) The number of BSs regarding the capacity is then the ceiling calculation of the number of users times the capacity of each user, divided by the capacity of each BS, divided by the number of sectors of each BS. This procedure can be summarised for each zone as follows:

1. The maximum throughput of a BS is calculated using the maximum MCS;
2. The utilisation factor (dependent on the geographical area) is multiplied by the maximum capacity, which maps the inter-cell interference;
3. The number of active users per zone is multiplied by the highest throughput considered for each user (1 Mbps in DL); (In capacity analysis, the DL will limit the number of BSs because the higher throughputs are provided in the BS-user direction.)
4. Then the result of Step 3 is divided (ceiled) by the result of Step 2 to get the number of required sectors;
5. Since each BS has three sectors, the previous result is again divided (ceiled) by three, to get the number of BSs.

For network planning, the region under study was divided into zones and the number of BSs was determined for each of them taking the population density into consideration. However, and mainly in rural areas, it can happen that the number of calculated BSs is higher than required. With this approach, at least one BS is required for each zone, but it could happen that a neighbour BS has enough available capacity to accommodate more users and, therefore a BS becomes surplus.

### 8.3.3 Required Number of BSs

The coverage and the capacity planning are of essential importance to the whole radio network planning. The coverage planning determines the service range, while the capacity planning determines the number of BSs to be used and their respective capacities. In this way, the main outcome of the network planning is the number of BSs that will provide coverage and at the same time fulfil the projected quality objectives for the different services offered. Therefore, the total number of BSs is calculated as the highest value between the coverage and capacity calculations, as detailed in Section 8.3.2.

For example, Table 8.3 shows the different number of calculated BSs for the legacy and TVWS bands in Munich (centre) for the first year. Clearly, the required number of BSs for frequency 2.6 GHz is higher than in 700 MHz, which in principle would have the advantage of reducing both CAPEX and OPEX.

### 8.3.4 Core and Transport Network Dimensioning

In addition to the number of BSs, the network topology must also be considered. As previously mentioned, the interfaces X2/S1 (see Fig. 8.4) are implemented

**TABLE 8.3** Number of BSs for 2.6 GHz and 700 MHz in Munich (Center) for the First Year

| Region | Zone | Number of users[a] | Number of BSs | |
|---|---|---|---|---|
| | | | 2.6 GHz | 700 MHz |
| Munich (center) | Altstadt-Lehelv (1) | 170 | 6 | 5 |
| | Ludwigsvorstadt-Isarvorstadt (2) | 412 | 14 | 12 |
| | Maxvorstadt (3) | 415 | 15 | 12 |
| | Schwabing-West (4) | 536 | 19 | 15 |
| | Au-Haidhausen (5) | 489 | 17 | 14 |
| | Sendling (6) | 334 | 12 | 10 |
| | Sendling-Westpark (7) | 458 | 16 | 13 |
| | Schwanthalerhöhe (8) | 235 | 8 | 7 |
| | Neuhausen-Nymphenburg (9) | 761 | 26 | 21 |
| | Moosach (10) | 430 | 15 | 12 |
| | Milbertshofen-Am Hart (11) | 603 | 21 | 17 |
| | Schwabing-Freimann (12) | 562 | 19 | 16 |
| | Bogenhausen (13) | 681 | 24 | 19 |
| | Berg am Laim (14) | 351 | 12 | 10 |
| | Trudering-Riem (15) | 485 | 17 | 14 |
| | Ramersdorf-Perlach (16) | 924 | 32 | 26 |
| | Obergiesing (17) | 423 | 15 | 12 |
| | Untergiesing-Harlaching (18) | 433 | 15 | 12 |
| | Thalkirchen-Obersendling-Forstenried-Fürstenried-Solln (19) | 726 | 25 | 20 |
| | Hadern (20) | 405 | 14 | 12 |
| | Pasing-Obermenzing (21) | 574 | 20 | 16 |
| | Aubing-Lochhausen-Langwied (22) | 341 | 12 | 10 |
| | Allach-Untermenzing (23) | 250 | 9 | 7 |
| | Feldmoching-Hasenbergl (24) | 488 | 17 | 14 |
| | Laim (25) | 451 | 16 | 13 |
| Total | | 11936 | 416 | 339 |

[a] *The same information as in 7th column of Table 8.1.*

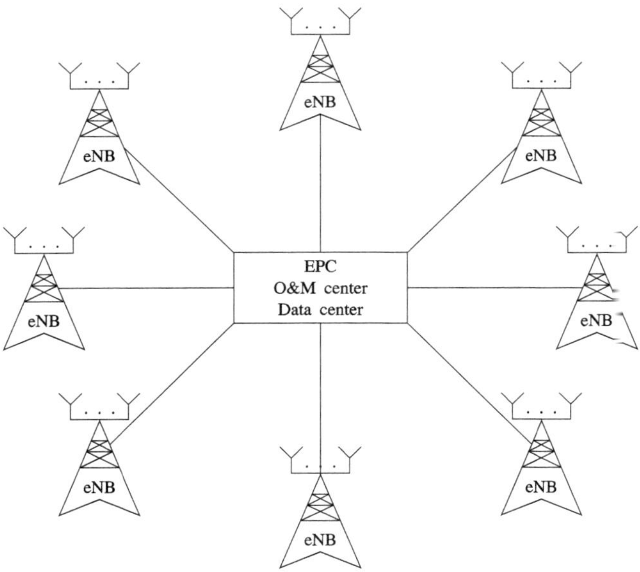


**FIGURE 8.5** EPC area with star topology for the network under evaluation.

through the core network, and data and control information can both be multiplexed in the same optical link. This is only true if the delay in the optical fibre can be neglected (when compared to the requirements of most demanding services, e.g. real-time services) and there is enough capacity in a single fibre to deliver all the traffic from the BS that is connected to the core network, and vice versa. In fact, the area under study is a small area of 50 × 50 km around a European metropolis, so the previous statement is expected to be true.

The total square area for evaluation can be divided into smaller squares. Each of these squares is named as an EPC area and represents an area where all the BSs are connected to that particular EPC, as in Fig. 8.5. Recall that EPCs can be limited by the number of BSs connected to a single MME or the traffic that can be processed at the same time by the S-GW and P-GW. Here, those limits are adjusted from [42] and equal to 4000 BSs or 40 Gbps, which are aligned (same order of magnitude) with the values used in [43].

## 8.4 DEPLOYMENT COSTS

In Section 8.3 we determined the number of network elements needed to fulfil coverage, capacity and service quality objectives. In this section, we estimate

**TABLE 8.4 Reference Values for Traditional Spectrum Market Deployments at Legacy and TVWS Carrier Frequencies**

| Parameter | Value | Ref. |
|---|---|---|
| Period of analysis (*N*) | 5 years (y0 to y4) | |
| Clients base (y0) | 18% (397 878 people) | [44] |
| Clients growth (y1 to y4)[a] | 80%, 128%, 179%, 215% | [44] |
| Price of devices (y0 to y4) | €80, €60, €40, €30, €20 | |
| Device subsidies | 50% | [45] |
| ARPU | €13 | |
| Price of spectrum (2.6 GHz, 700 MHz)[b] | €3 million, €45 million | [46] |
| Price of optical fiber | €45 000/km | [45] |
| Fiber sharing factor[c] | 50% | [45] |
| Price of EPC | €50 000 | [45] |
| Price of BS | €3500 | [45] |
| Price of O&M and data centers[d] | €500 000 | |
| Interest rate | 5% | |

[a] *In comparison with the first year of deployment (y0).*
[b] *For a bandwidth pair of 2 × 5 MHz. Also, despite of having three sectors per cell, only a carrier is paid, i.e., the same carrier is reused in all sectors.*
[c] *In the optical fiber, control information and data are multiplexed. This percentage refers to the amount of fiber used for data.*
[d] *Information obtained from a Brazilian operator in a private document.*

the cost of each considered element. Some of the values (e.g. cost of traditional spectrum market, price of devices or interest rate) are based on real values, while others (such as cost of optical fibre infrastructure, EPC and BS) are found in the literature [44,45].

Table 8.4 summarises the most relevant parameters and their respective values for deployment scenarios 1 and 2, while Table 8.5 has the additional inputs for the deployment using the legacy carrier plus a TVWS carrier, i.e. deployment scenario 3. In both cases, these values are the references used to calculate the total NPV and to start the sensitivity analysis.

To complement the values shown in Tables 8.4 and 8.5, the other costs associated with the infrastructure of EPC and BSs, and O&M upgrades (hardware or software) are also considered. Finally, the price of the TVWS carrier per day in the context of flexible spectrum market and its average use during the day (considering every day of the year) reflects the operator's necessity to overcome capacity faults, especially found during busy hours. The cost of spectrum is based on the Portuguese spectrum auctions [46].

**TABLE 8.5 Reference Values for Flexible Spectrum Market Deployment Using the Legacy Carrier Plus an Additional TVWS Carrier**

| Parameter | Value | Ref. |
|---|---|---|
| Price of spectrum (2.6 GHz) | €3 million | [46] |
| Price of TVWS/day (urban, sub-urban, rural)[a] | €411, €247, €164 | [24] |
| BSs that have additional TVWS carrier (urban, sub-urban, rural)[b] | 50%, 30%, 20% | |
| Average use of TVWS during the day | 4 hours | [24] |
| ARPU (increased by 10%) | €14.3 | |
| Price of BS (increased by 10%) | €3850 | [45] |

[a] *As before, for a bandwidth pair of 2 × 5 MHz.*
[b] *Values are based on the expectation that higher densely populated zones require more BSs with additional TVWS carrier.*

## 8.5 RESULTS AND DISCUSSION

In this section we present the results of the three different deployment scenarios, as introduced in Section 8.1:

- Deployment scenario 1: traditional spectrum market for deployment using only the legacy carrier of 2.6 GHz;
- Deployment scenario 2: traditional spectrum market for deployment using only the TVWS carrier of 700 MHz;
- Deployment scenario 3: flexible spectrum market for deployment using the legacy carrier plus an additional TVWS carrier.

The analysis is divided into the following aspects: NPV, costs (CAPEX and OPEX), and the sensitivity of NPV to the change in an input parameter. The basis for the whole analysis presented here employs the values from Tables 8.4 and 8.5, while Eq. (8.3) is used for the NPV calculation.

### 8.5.1 NPV, CAPEX and OPEX for Deployment Scenarios 1 and 2

In Fig. 8.6 we present the total NPV and the balance or cash flow (revenues minus costs) per year as in Eq. (8.1), for the total of five years of deployment. As can be seen for the first year, which is the year when most of the network is implemented, the balance is negative; it is more negative for the carrier frequency of 2.6 GHz than 700 MHz. For the other years, the balance is positive, which means that the total costs are less than the revenues in each particular year.

Regarding the contribution of each geographical area to the total NPV, the largest contribution comes from urban and suburban areas (see Fig. 8.7). This result was expected because in the square under evaluation the network is mainly

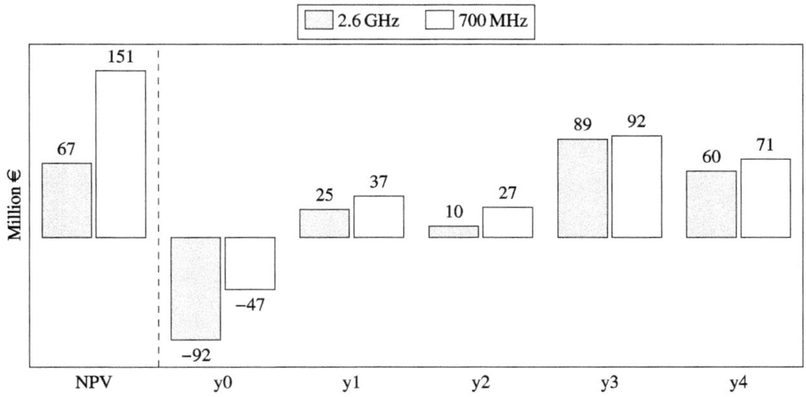


FIGURE 8.6 Total NPV and balance per year.

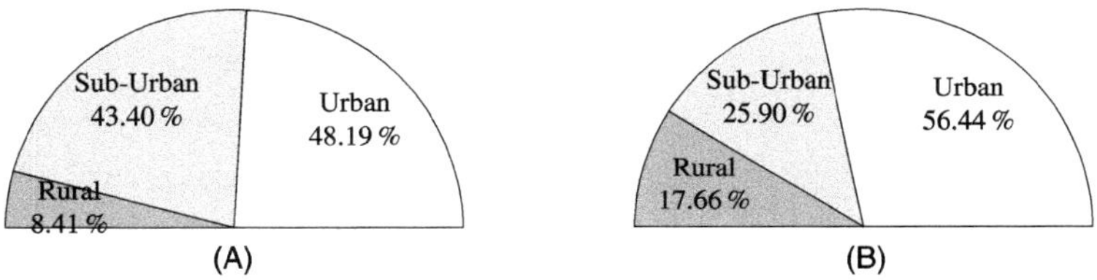


FIGURE 8.7 Contribution of each geographical area for the total NPV. (A) 2.6 GHz; (B) 700 MHz.

limited, due to capacity, and most of the people are located in urban and suburban areas.

Looking at the cumulative figures of costs and revenues in Fig. 8.8(A), if we draw a line between the edges of the bars for the deployment using 2.6 GHz and, in the same manner, for the bars of 700 MHz (see Fig. 8.8(B)), it is easy to see that the 2.6 GHz line is above the 700 MHz line, which means that the costs are greater and, comparing the slopes, tend to increase more rapidly.

Moreover, if we do the same exercise for the revenues, we can see that its line crosses the 700 MHz line around the second year and the 2.6 GHz line between the third and fourth years. Therefore, it is clear that an operator would recover its investment one and half years earlier if it chose to deploy the network in 700 MHz rather than in 2.6 GHz. Also, the slope of the revenues line is greater than that of any of the cost lines, i.e. the revenues increase faster than the costs; and in both deployments the network will be profitable.

Furthermore, it is useful to determine the main contributions of the deployment and operational costs, which are presented in Fig. 8.9 for CAPEX and in Fig. 8.10 for OPEX. Regarding CAPEX, and contrary to what was expected, the impact of spectrum costs is not significant (only 0.27% for 2.6 GHz and 5.51%

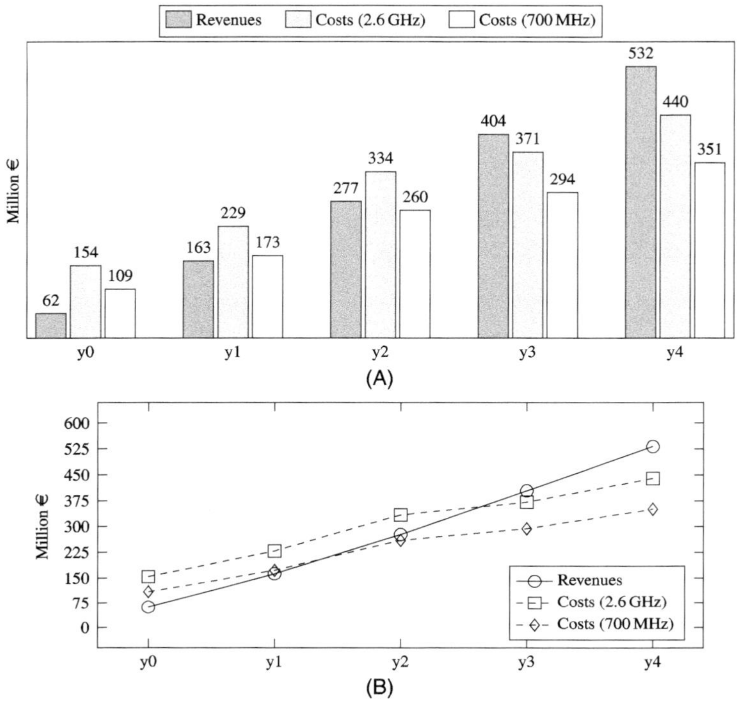


**FIGURE 8.8** Cumulative total revenues and costs per year. (A) Bar chart; (B) Transition points.

for 700 MHz). On the other hand, the cost of fibre represents around 90% of the CAPEX. Thus, we can reduce CAPEX if we are able to reduce the size of the transport network. Regarding OPEX, the costs related to the management of optical fibre infrastructure are once again revealed to be significant (around 28% for 2.6 GHz and 23% for 700 MHz); however, the costs with subsidies and replacement of devices represent the biggest portion of the OPEX (around 62% for 2.6 GHz and 69% for 700 MHz).

The previous results seem to give advantage to the 700 MHz frequency. However, the results must be assessed regarding their sensitivity to the change in the input parameters, as is presented in the following paragraphs.

## 8.5.2 Sensitivity Analysis for Deployment Scenarios 1 and 2

The sensitivity analysis is performed in the following manner: the value of an input parameter is changed while the others remain the same and then the value of NPV is extracted. This value is the sensitivity of NPV to the change (increase or decrease) in that parameter. The sensitivity analysis is useful in order to de-

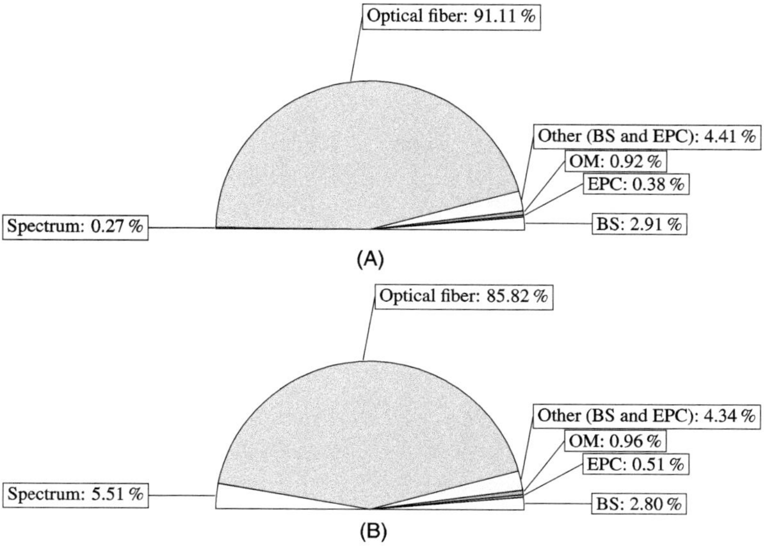


**FIGURE 8.9** Contribution of each parameter to total CAPEX. (A) 2.6 GHz; (B) 700 MHz.

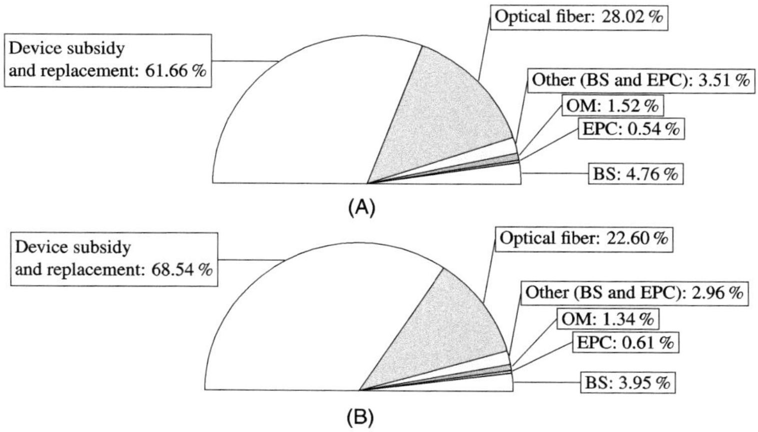


**FIGURE 8.10** Contribution of each parameter to total OPEX. (A) 2.6 GHz; (B) 700 MHz

termine the influence of each parameter in the total NPV and then to establish strategies of deployment to mitigate or benefit from such influence.

In Fig. 8.11 the variation in NPV considering a reduction in client base is presented. The reference (REF) values from Table 8.4 are 397 878 people in y0 (i.e. the first year or year of implementation), 716 187 people in y1, 907 170 people in y2, 1 110 883 people in y3, and 1 253 480 people in y4. The slope of

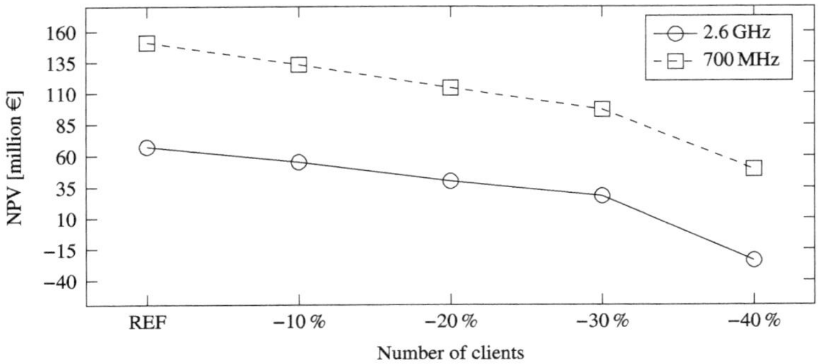


**FIGURE 8.11** Total NPV variation by reducing the number of clients.

both lines is high, which means that NPV is very sensitive to the reduction in the client base.

The higher slope presented in the reduction from 30 to 40% in the client base is derived from the fact that with fewer clients the operator receives fewer revenues, but (as explained in Section 8.3.2) if there is at least one client per zone, one BS (and EPC) needs to be deployed to serve that client; for a reduction of 40% in the client base the NPV in 2.6 GHz is negative, i.e. the network becomes non-profitable.

Fig. 8.12 presents the NPV variation while increasing the price of devices. The REF values are €80 in y0, €60 in y1, €40 in y2, €30 in y3 and €20 in y4. Since the slope of the lines is not very high, the NPV is not very sensitive to the increase in the price of devices. In a similar way, Fig. 8.13 shows the variation of the total NPV due to the increase in device subsidies that an operator gives to its clients. The REF value is 50%. As can be seen, the slope of both lines is not very high, thus the NPV is not very sensitive to the increase in the device subsides, but more sensitive than to the increase in price of devices.

Instead, it is clear that the NPV is highly sensitive to the reduction in ARPU, which can be observed in Fig. 8.14. Starting with a REF value of €13, the slope of the lines is very high and the NPV becomes negative if ARPU reduces by around 13% in the 2.6 GHz deployment scenario, while for 700 MHz the reduction must be around 32%. Nevertheless, in both cases the network becomes non-profitable. Moreover, this result shows that NPV is more sensitive to the reduction in ARPU than to the reduction in the client base.

Fig. 8.15 shows the NPV variation considering the increase in the price of spectrum. The REF is €3 million for 2.6 GHz and €45 million for 700 MHz. In this result, both lines appear to be nearly horizontal. This means that NPV is not sensitive to the increase in the price of spectrum. It may seem surprising for 700 MHz because (as shown in Fig. 8.9(B)) the price of spectrum represents

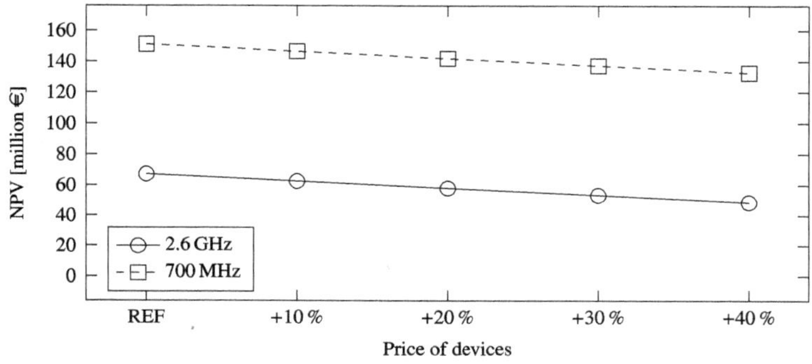


**FIGURE 8.12** Total NPV variation due to increasing the price of devices.

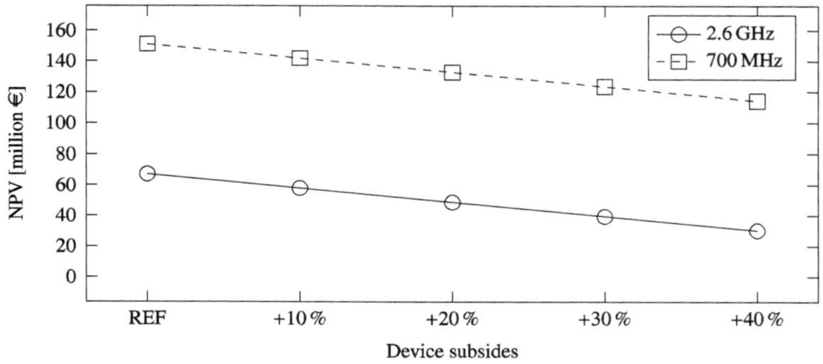


**FIGURE 8.13** Total NPV variation due to increasing the device subsides.

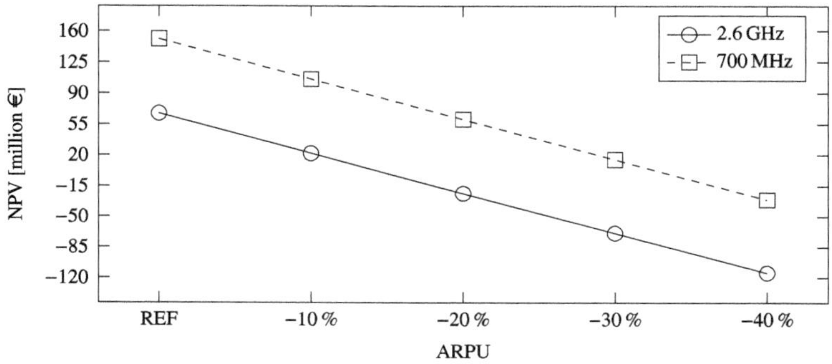


**FIGURE 8.14** Total NPV variation due to reducing the ARPU.

5.51% of the CAPEX for the reference values. However, if we analyse the impact of increasing the price of spectrum in CAPEX, we see that its impact in the worst case (70% increase) is only around 9% (see Fig. 8.16); this is a varia-

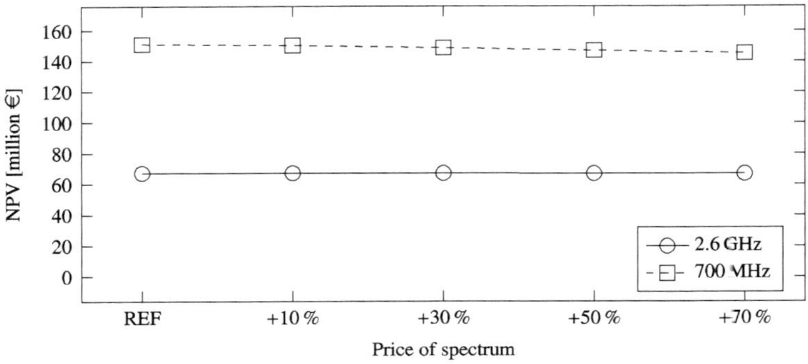


**FIGURE 8.15** Total NPV variation due to increasing the price of spectrum.

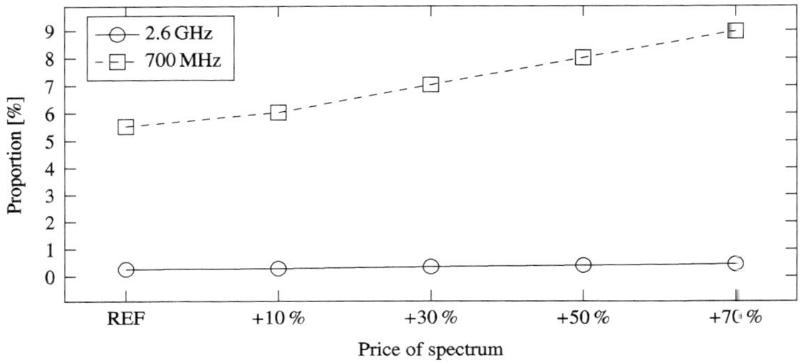


**FIGURE 8.16** Proportion of spectrum costs in the total CAPEX due to increasing the price of spectrum.

tion of only 3.49%, which demonstrates that its lesser impact is in line with the previous result.

For the transport network, Fig. 8.17 presents the variation in NPV considering the increase in the optical fibre costs. The REF value is €45 000/km for a new fibre installation. As can be seen, the increase in optical fibre costs has a greater impact on the deployment with 2.6 GHz than with 700 MHz. In deployment scenario 1 the radii of cells are smaller due to propagation loss effects. Therefore, to cover the same area, more BSs are required and more fibre is also required to connect the additional BSs, which explains the difference between the slopes of the two lines.

### 8.5.3 NPV, CAPEX and OPEX for Deployment Scenario 3

With the introduction of a TVWS carrier (700 MHz) in addition to the already planned network using the central carrier frequency of 2.6 GHz, some adjust-

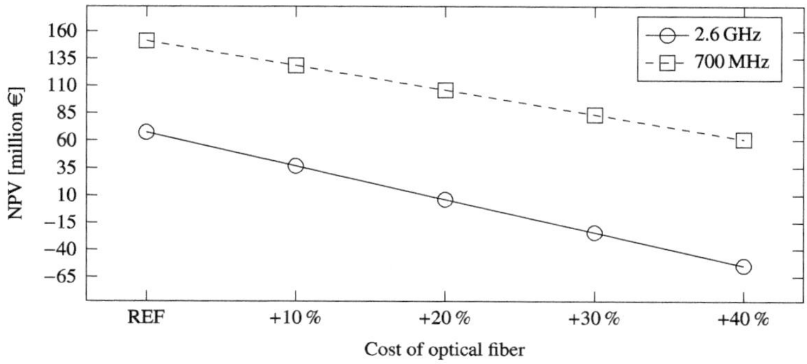


**FIGURE 8.17** Total NPV variation due to increasing the cost of optical fibre.

ments were necessary. These adjustments are shown in Table 8.5. Specifically, a cost of 10% for each BS was added because new antennas are needed for the new carrier. An additional cost for the TVWS carrier depending on the geographical classification was also introduced; in urban areas the demand for TVWS will be higher, which increases its price.

Moreover, and through system level simulations, it was possible to verify an improvement in the average throughput per user of around 8% in urban areas (1.08 Mbps), and 14% in suburban and rural areas (1.14 Mbps) [29]. In these cases, the operator can comfortably increase ARPU by 10% (€14.3) because a better Quality of Service (QoS) will be provided. It is also important to note that the extra capacity is expected to improve the throughput of best effort services because the network was planned with only a carrier of 2.6 GHz in order to guarantee the estimated traffic in DL of 1 Mbps and in UL of 256 kbps per user.

Fig. 8.18 presents the comparison between the NPV and balance per year of deploying the network when considering only the 2.6 GHz carrier and when the deployment is done using the 2.6 GHz plus an additional TVWS carrier; the deployments use the reference values presented in Tables 8.4 and 8.5, respectively.

In both situations the NPV is positive, which means that the project is profitable. Regarding the balance, it is only negative in the first year (y0). As discussed before, the first year is when most of the investment is made, so a negative value is normal, and not that much different in the two deployment scenarios. However, looking at other years of the study, the balance per year is more positive for 2.6 GHz plus TVWS than when considering a deployment with only the 2.6 GHz carrier (the only exception is y3); therefore, the total NPV is clearly higher. This result is in fact expected because, as already seen in Fig. 8.14, the NPV is highly sensitive to changes in ARPU, thus even with a small increase in ARPU (10%) for deployment scenario 3, the effect on NPV is high.

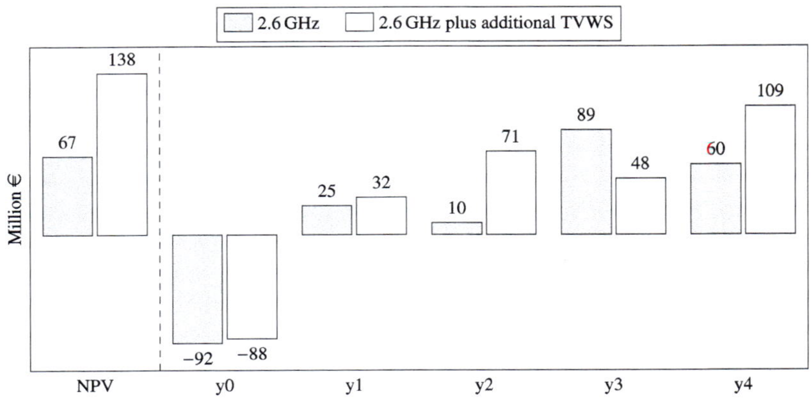


**FIGURE 8.18** Total NPV and balance per year.

The cumulative revenues and costs per year for deployment scenarios 1 and 3 are shown in Figs 8.19 and 8.20, respectively. If we look exclusively at Figs 8.19(A) and 8.20(A) we might be tempted to say that the difference between the two deployments considering both the revenues and costs is not significant. However, if we look at the third year (y2) we see that only in the case of deployment for 2.6 GHz plus an additional TVWS is the revenues bar slightly above the costs bar for that year; in the case of deployment with 2.6 GHz only, the revenues bar is clearly above the costs bar only in the fourth year (y3).

A better visual illustration of the same result is depicted in Fig. 8.20(B) where the revenues line crosses the costs line before the third year of deployment, while in Fig. 8.19(B) it happens almost at the end of the third year. In other words, it is clear that an operator would recover its investment around one year earlier if it chose to deploy the network following a flexible spectrum market approach.

Despite the previous results, it is a good idea to determine the main sources of costs in the network. In Fig. 8.21 the principal contributors to CAPEX and OPEX for deployment scenario 3 can be seen. Regarding the CAPEX, it is very similar to Fig. 8.9(A), which was done for a network with only the 2.6 GHz carrier. Actually, this result was expected because the only added cost to CAPEX is the increase of 10% in the cost of each BS due to the required additional antennas to support TVWS. The main difference can be seen in OPEX where the annual price of spectrum for TVWS now represents around 12% of the total OPEX. Nevertheless, device subsidies and replacement costs are still the most important contributors to OPEX.

Notwithstanding the big impact on OPEX, the price of TVWS does not have a visible impact on the total NPV (which remains very positive). This can be

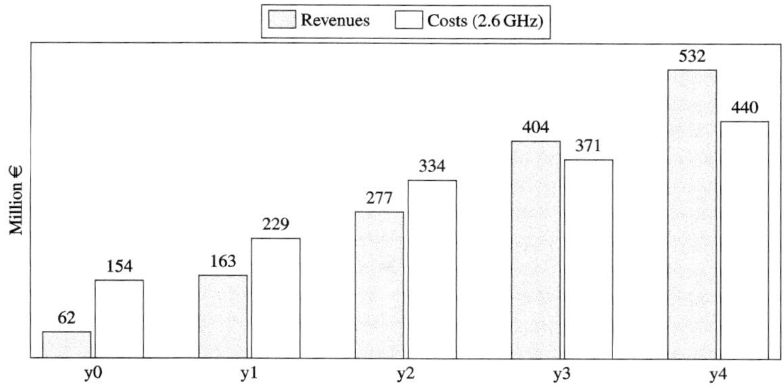


(A)

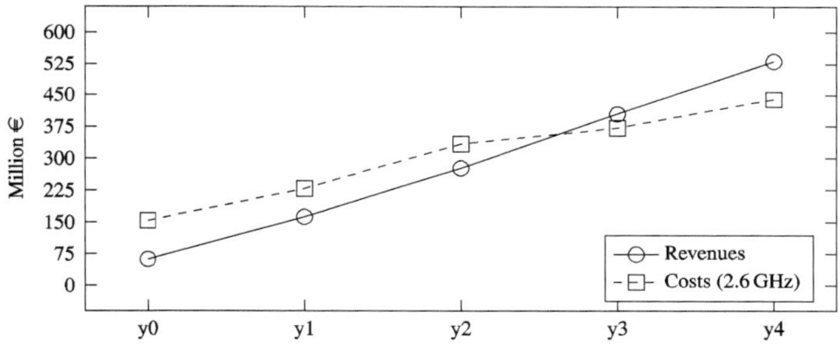


(B)

**FIGURE 8.19** Cumulative total revenues and costs per year for deployment scenario 1. (A) Bar chart; (B) Transition points.

explained with the reference values: 81% of the costs are CAPEX while only 19% are OPEX.

## 8.5.4 Sensitivity Analysis for Deployment Scenario 3

In addition to the sensitivity analysis already performed in Section 8.5.2, here we add two more sensitivity analyses to assess the influence on the total NPV of increasing the price of a TVWS carrier and the number of hours per day (on average) that an operator must pay to use this carrier (or TVWS channel).

As such, Fig. 8.22 shows the variation in NPV while the price of TVWS increases. The REF values are €411/day for urban, €247/day for suburban, and €164/day for rural areas. In both cases, for the average use of four hours/day or eight hours/day the slope of the lines is not very steep; in other words, the NPV is not very sensitive to the increase in the price of TVWS.

To finalise the sensitivity analysis, Fig. 8.23 shows the variation in NPV while increasing the number of hours/day that an operator must pay for the

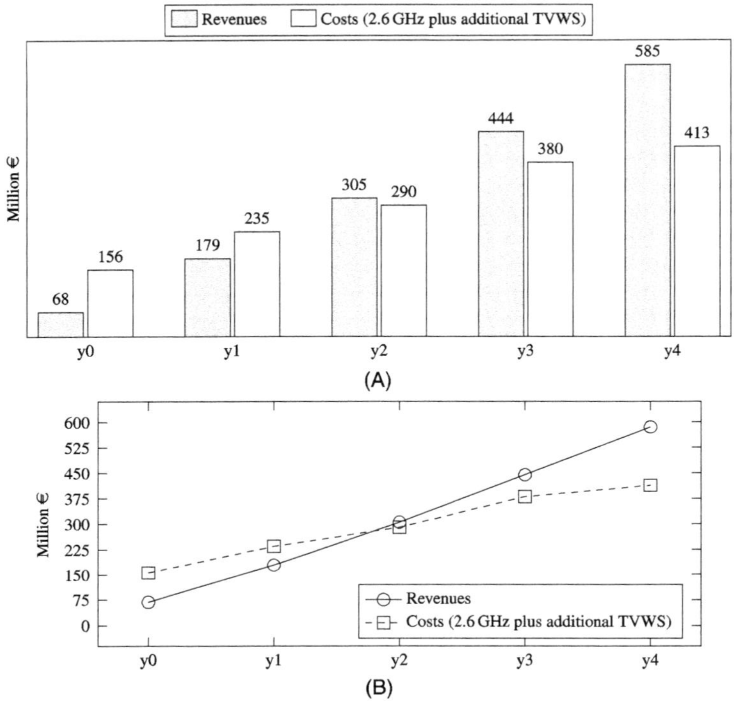


**FIGURE 8.20** Cumulative total revenues and costs per year for deployment scenario 3. (A) Bar chart; (B) Transition points.

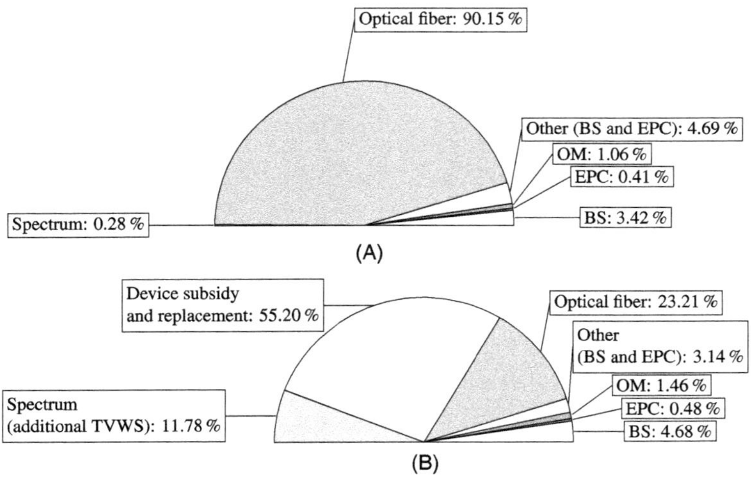


**FIGURE 8.21** Contribution for the total costs of each parameter. (A) CAPEX; (B) OPEX.

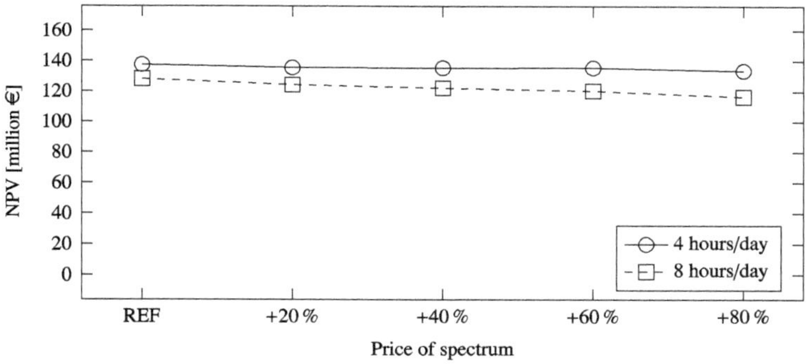


**FIGURE 8.22** Total NPV variation due to increasing the price of spectrum (additional TVWS).

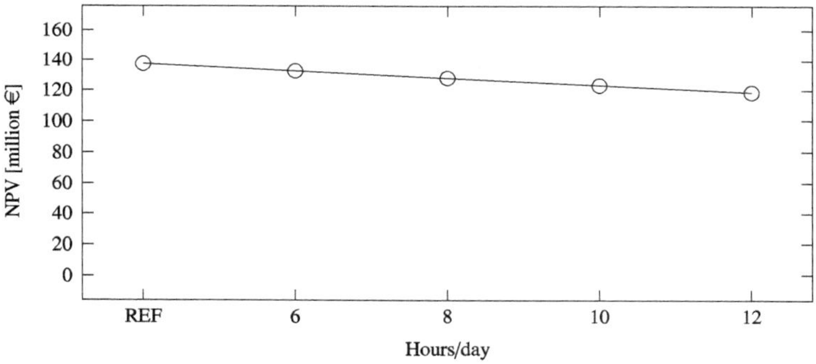


**FIGURE 8.23** Total NPV variation due to increasing the number of paid TVWS hours.

use of a TVWS carrier. The REF value is four hours/day, which represents on average the typical use time per day. As can be seen, once again, the variation in NPV is not very high (around €132 million for REF to around €113 million for 12 hours/day).

Since using TVWS is meant to overcome situations where there is a lack of capacity during busy hours, it does not make sense to use the TVWS during the whole day. However, even for the worst case, that is when the operator must pay 24 hours/day for the use of a TVWS carrier, the value of NPV is €84 million, which is clearly above the €64 million obtained when considering only a carrier of 2.6 GHz.

## 8.6 CONCLUSION

The techno-economic evaluation is a way of investigating the profitability of a project and, if correctly done, can help decision makers to implement structured strategies. However, it is almost impossible to assess all the components and their contributions, even more difficult when the study is for some time in future, because of the time-dependency of many variables (e.g. prices, interest rates or the inflation rate). It may be even worse for something that is completely new because there is no past knowledge that can help to predict the future. Therefore, some assumptions of values and their variation over time must be made.

The NPV metric is commonly used to assess the profitability of a project: a positive NPV shows that the project adds value to the company and should be sustained, while a negative NPV means that the project takes value from the company and, accordingly, should be abandoned; if NPV is zero, other metrics (or subjective factors) should be used instead.

The study described in this work is for the region of Munich, and considers three different LTE network deployments: with only one carrier, either 2.6 GHz (scenario 1) or 700 MHz (scenario 2), for the traditional spectrum market and a with the legacy carrier of 2.6 GHz plus an additional TVWS carrier (scenario 3), which is aligned with a flexible spectrum market approach.

Using reference values from the real market and literature, it was shown that the network is clearly more profitable in deployment scenario 2 where the NPV reaches €151 million, while in deployment scenario 3 it reaches €138 million, and for deployment scenario 1 the NPV is only €67 million. Therefore, it is evident that an operator will recover its investment one and half years earlier if it chooses to deploy the network in 700 MHz rather than in 2.60 GHz, and around one year earlier if it chooses to deploy the network following a flexible spectrum market approach with an additional TVWS carrier.

Nevertheless, absolute values of NPV should be taken with caution due to the assumptions that are made. Therefore, a sensitivity analysis is highly beneficial because it gives a broader view of the factors that influence the NPV and, as a result, the profitability of the project. It is also very useful to establish strategies, e.g. investing in a certain factor that has less influence on the NPV but beneficially interferes with other factors that highly affect the NPV.

Table 8.6 presents a summary of the sensitivity of NPV to the variation in each input parameter in the three deployment scenarios (a visual illustration of the same result is presented in Fig. 8.24). As can be seen, the ARPU is the most important contributor to the total NPV, followed by the cost of fibre, and the client base; instead, the NPV is less sensitive to the variations in the price of devices, cost of spectrum (traditional market or additional TVWS carrier), and the use time of TVWS.

With the combination of NPV and sensitivity analysis for deployment with a legacy carrier of 2.6 GHz or when an additional TVWS carrier is added to the legacy frequency, it becomes clear that with the introduction of TVWS, following a flexible spectrum market approach, the level of profitability increases and does not vary much; in other words, it is not too sensitive to changes in the price

**TABLE 8.6 Summary of the Sensitivity Analysis Regarding the Variation of Parameters**

| Parameter | Sensitiveness of NPV | |
|---|---|---|
| | *Scenarios 1 and 3* | *Scenario 2* |
| Clients base | Moderate | Moderate |
| Price of devices | Small | Small |
| Device subsidies | Small | Small |
| ARPU | Very high | Very high |
| Price of spectrum (traditional market) | Very small | Very small |
| Price of optical fiber | High | Moderate |
| Price of spectrum (additional TVWS per 4 hours/day)[a] | Very small | |
| Price of spectrum (additional TVWS per 8 hours/day)[a] | Very small | |
| Use time of TVWS[a] | Very small | |

[a] *Only for the deployment scenario 3.*

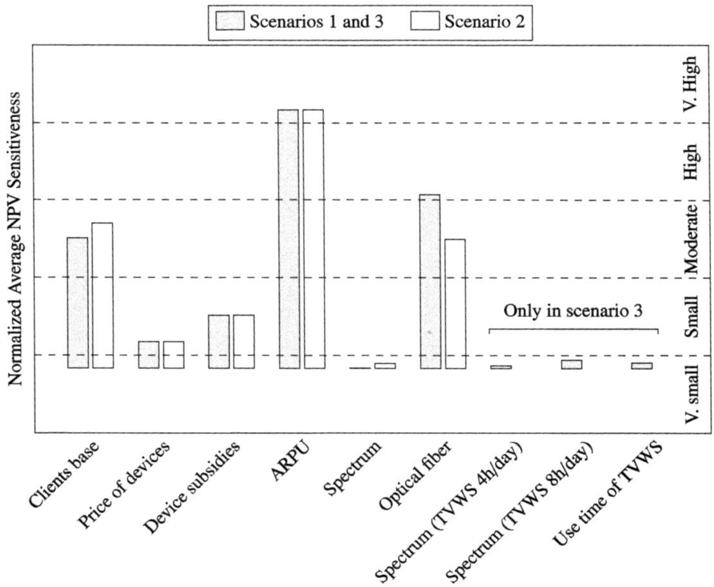


**FIGURE 8.24** Normalised average NPV sensitivity, which is the slope of the total NPV line while varying each parameter, divided by the highest achieved slope of the total NPV line among all parameters.

or use time of TVWS. In this case, the operator may feel comfortable to slightly increase the ARPU, which highly influences the NPV, because with TVWS the network capacity increases and thus a better quality QoS can be provided.

As another example, it was noted in [25] that one of the threats is the expected high price of WSDs. An incentive to the clients is to increase device subsidies or to decrease the device prices, for example with promotions to attract new clients and increase the client base. The impact of the client base on the NPV is moderate, while the effect of device subsidies and the price of devices is small.

A more formal mathematical approach for the sensitivity analysis with the use of partial derivatives would also be beneficial since it would be possible to clearly identify the (marginal) contribution of each parameter to the total NPV. However, there is no way to isolate these parameters within Eq. (8.3) and thus to have a closed-form formula to calculate NPV, which means that this approach is impractical.

Finally, future works may involve the deployment of optimisation strategies to reduce the network costs while keeping the same coverage and capacity, and evaluating the use of TVWS in that context. Additionally, the use of NPV with Monte Carlo runs [27,47] to simulate uncertainties with a statistical characteristic (i.e. statistical distribution) is useful to model the risks associated with the flexible spectrum market, and to perform the sensitivity analysis.

## REFERENCES

[1] Cisco, Cisco visual networking index: global mobile data traffic forecast update, 2014–2019, White paper, Feb. 2015 [Online]. Available: http://www.cisco.com/c/en/us/solutions/collateral/service-provider/visual-networking-index-vni/white_paper_c11-520862.pdf.

[2] Ericsson, Ericsson mobility report, Jun. 2016 [Online]. Available: https://www.ericsson.com/res/docs/2016/ericsson-mobility-report-2016.pdf.

[3] J.G. Andrews, S. Buzzi, W. Choi, S.V. Hanly, A. Lozano, A.C.K. Soong, J.C. Zhang, What will 5G be?, IEEE Journal on Selected Areas in Communications 32 (6) (Jun. 2014) 1065–1082.

[4] A. Gupta, R.K. Jha, A survey of 5G network: architecture and emerging technologies, IEEE Access 3 (Jul. 2015) 1206–1232.

[5] ITU, Draft new report ITU-R M.[IMT-2020.TECH PERF REQ] – minimum requirements related to technical performance for IMT-2020 radio interface(s), Feb. 2017 [Online]. Available: https://www.itu.int/dms_pub/itu-r/md/15/sg05/c/R15-SG05-C-0040!!MSW-E.docx.

[6] 3GPP, Technical specification group radio access network; study on scenarios and requirements for next generation access technologies, 3rd Generation Partnership Project (3GPP), TR 38.913, Mar. 2017, v14.2.0 [Online]. Available: http://www.3gpp.org/ftp/specs/html-info/38913.htm.

[7] E. Hossain, M. Rasti, H. Tabassum, A. Abdelnasser, Evolution toward 5G multi-tier cellular wireless networks: an interference management perspective, IEEE Wireless Communications 21 (3) (Jun. 2014) 118–127.

[8] E.G. Larsson, F. Tufvesson, O. Edfors, T.L. Marzetta, Massive MIMO for next generation wireless systems, IEEE Communications Magazine 52 (2) (Feb. 2014) 186–195.

[9] S. Rangan, T.S. Rappaport, E. Erkip, Millimeter wave cellular wireless networks: potentials and challenges, Proceedings of the IEEE 102 (3) (Feb. 2014) 366–385.
[10] S. Mumtaz, K.M.S. Huq, J. Rodriguez, Direct mobile-to-mobile communication: paradigm for 5G, IEEE Wireless Communications 21 (5) (Oct. 2014) 14–23.
[11] L. Wei, R.Q. Hu, Y. Qian, G. Wu, Enable device-to-device communications underlaying cellular networks: challenges and research aspects, IEEE Communications Magazine 52 (6) (Jun. 2014) 90–96.
[12] J.G. Andrews, Seven ways that HetNets are a cellular paradigm shift, IEEE Communications Magazine 51 (3) (Mar. 2013) 136–144.
[13] J. Mitola III, et al., Accelerating 5G QoE via public-private spectrum sharing, IEEE Communications Magazine 52 (5) (May 2014) 77–85.
[14] T.M. Taher, R.B. Bacchus, K.J. Zdunek, D.A. Roberson, Long-term spectral occupancy findings in Chicago, in: 2011 IEEE Symposium on Dynamic Spectrum Access Networks (DySpan), May 2011, pp. 100–107.
[15] A. Palaios, J. Riihijärvi, O. Holland, P. Mähönen, A week in London: spectrum usage in metropolitan London, in: IEEE 24th Annual International Symposium on Personal, Indoor, and Mobile Radio Communications (PIMRC), Sep. 2013, pp. 2532–2537.
[16] M. Prazy, The profitability analysis of the multi-band spectrum broker, in: IEEE 24th Annual International Symposium on Personal, Indoor, and Mobile Radio Communications (PIMRC), Sep. 2013, pp. 3454–3458.
[17] OFCOM, Digital dividend review: geographic interleaved awards 470–550 MHz and 630–790 MHz, Consultation (Jun. 2008).
[18] E.C.C. (ECC), Technical considerations regarding harmonisation options for the digital dividend, in: European Conference of Postal and Telecommunications Administrations (CEPT), in: CEPT Report, vol. 21, Mar. 2007 [Online]. Available: www.erodocdb.dk/docs/doc98/official/pdf/CEPTREP021.pdf.
[19] E.C.C. (ECC), Technical considerations regarding harmonisation options for the digital dividend, in: European Conference of Postal and Telecommunications Administrations (CEPT), in: CEPT Report, vol. 24, Jun. 2008 [Online]. Available: http://www.erodocdb.dk/docs/doc98/official/pdf/CEPTRep024.pdf.
[20] COGEU, European TV white spaces analysis and COGEU use-cases, Deliverable 2 (1) (Mar. 2010), version 1.0 [Online]. Available: http://www.ict-cogeu.eu/deliverables.html.
[21] ITU-R, Definitions of software defined radio (SDR) and cognitive radio system (CRS), Report ITU-R SM.2152, Sep. 2009 [Online]. Available: https://www.itu.int/dms_pub/itu-r/opb/rep/R-REP-SM.2152-2009-PDF-E.pdf.
[22] FCC, In the matter of unlicensed operation in the TV broadcast bands: Et docket no. 04-186, Washington D.C., Order DA 11-131, Jan. 2011.
[23] OFCOM, Implementing TV white spaces, Statement, Feb. 2015 [Online]. Available: http://stakeholders.ofcom.org.uk/binaries/consultations/white-space-coexistence/statement/tvws-statement.pdf.
[24] H. Bogucka, M. Parzy, P. Marques, J.W. Mwangoka, T. Forde, Secondary spectrum trading in TV white spaces, IEEE Communications Magazine 50 (11) (Nov. 2012) 121–129.
[25] C.F.M. Silva, F.R.P. Cavalcanti, Á. Gomes, SWOT analysis for TV white spaces, Transactions on Emerging Telecommunications Technologies 26 (6) (Jun. 2015) 957–974, published online: Dec. 2013.
[26] S. Bichler, J. Nitzan, Systemic fear, modern finance and the future of capitalism, Jerusalem and Montreal, Jul. 2010 [Online]. Available: http://bnarchives.yorku.ca/289/03/20100700_bn_systemic_fear_modern_finance_future_of_capitalism.pdf.
[27] S.C. Savvides, Risk analysis in investment appraisal, Project Appraisal 9 (1) (Mar. 1994) 3–18 [Online]. Available: https://mpra.ub.uni-muenchen.de/10035/1/MPRA_paper_10035.pdf.

[28] V. Maroyi, Capital Budgeting Practices: a South African Perspective, Master's thesis, Wageningen University, Aug. 2011.

[29] COGEU, Spectrum-aware routing, transport protocols and negotiation protocols between players for secondary spectrum trading; system level simulation tool – initial specification, Deliverable 6.2, Dec. 2011, version 1.0 [Online]. Available: http://www.ict-cogeu.eu/deliverables.html.

[30] F.J. Gallego, Mapping rural/urban areas from population density grids, Institute for Environment and Sustainability, JRC, 2004 [Online]. Available: http://www.ec-gis.org/docs/F11116/RURAL%20URBAN%20%20POPDENS.PDF.

[31] L. Dijkstra, H. Poelman, A harmonised definition of cities and rural areas: the new degree of urbanisation, European Commission, Working Papers 01/2014, May 2014 [Online]. Available: http://ec.europa.eu/regional_policy/en/information/publications/working-papers/2014/a-harmonised-definition-of-cities-and-rural-areas-the-new-degree-of-urbanisation.

[32] E. Dahlman, S. Parkvall, J. Sköld, 4G: LTE/LTE-Advanced for Mobile Broadband, 1st ed., Elsevier, 2011.

[33] H.R.R. Alves, Sistemas Com Rádios Cognitivos Para a Partilha Eficiente Dos Espaços Vazios da TV Com LTE [Systems with Cognitive Radios for an Efficient Sharing Between TV White Spaces and LTE], Master's thesis, Universidade da Beira Interior, Oct. 2011, original document in Portuguese [Online]. Available: http://hdl.handle.net/10400.6/3821.

[34] H. Holma, A. Toskala, LTE for UMTS: OFDMA and SC-FDMA Based Radio Access, 1st ed., John Wiley & Sons, 2009.

[35] V. Erceg, L.J. Greenstein, S.Y. Tjandra, S.R. Parkoff, A. Gupta, B. Kulic, A.A. Julius, R. Bianchi, An empirically based path loss model for wireless channels in suburban environments, IEEE Journal on Selected Areas in Communications 17 (7) (Jul. 1999) 1205–1211.

[36] V. Erceg, et al., Channel models for fixed wireless applications, IEEE 802.16 Broadband Wireless Access Working Group, IEEE 802.16.3c-01/29r4, Jul. 2001 [Online]. Available: http://www.ieee802.org/16/tg3/contrib/802163c-01_29r4.pdf.

[37] M. Hata, Empirical formula for propagation loss in land mobile radio services, IEEE Transactions on Vehicular Technology 29 (3) (Aug. 1980) 317–325.

[38] A.B. Syed, Dimensioning of LTE Network. Description of Models and Tools, Coverage and Capacity Estimation of 3GPP Long Term Evolution, Master's thesis, Helsinki University of Technology, Feb. 2009.

[39] P. Vieira, P. Queluz, A. Rodrigues, LTE spectral efficiency using spatial multiplexing MIMO for macro-cells, in: 2nd International Conference on Signal Processing and Communication Systems (ICSPCS'08), Dec. 2008, pp. 1–6.

[40] 3GPP, Evolved universal terrestrial radio access (E-UTRA), base station (BS) radio transmission and reception, 3rd Generation Partnership Project (3GPP), TS 36.104, Sep. 2012 [Online]. Available: http://www.3gpp.org/ftp/Specs/html-info/36104.htm.

[41] 3GPP, Evolved universal terrestrial radio access (E-UTRA); physical layer procedures, 3rd Generation Partnership Project (3GPP), TS 36.213, Mar. 2008 [Online]. Available: http://www.3gpp.org/ftp/specs/html-info/36213.htm.

[42] Motorola, Motorola WBC 700 portfolio, Jan. 2010.

[43] D. Dababneh, LTE Traffic Generation and Evolved Packet Core (EPC) Network Planning, Master's thesis, Carleton University, Mar. 2013.

[44] FUTON, Architecture, business evaluation and deployment models: final vision at the end of FUTON, Deliverable 2.3, Sep. 2010, version 1.2 [Online]. Available: http://www.ict-futon.eu/private/docs/deliverables/d2.3.pdf.

[45] C.F. Silva, V.S. Ribeiro, Business evaluation for FUTON-like architectures, in: IEEE International Conference on Computer as a Tool (EUROCON'11), Apr. 2011, pp. 1–4.

[46] ANACOM, Projecto de regulamento do leilão para a atribuição de direitos de utilização de frequências nas faixas dos 450 MHz, 800 MHz, 900 MHz, 1800 MHz, 2.1 GHz e 2.6 GHz [Regulation project for the attribution of usage rights of frequencies in the bands of 450 MHz, 800 MHz, 900 MHz, 1800 MHz, 2.1 GHz and 2.6 GHz], Mar. 2011, Regulamento, original document in Portuguese.
[47] M.v. Bredow, G. Reinhart, Method to evaluate the profitability and risk structure of projects in the automotive industry, in: IEEE International Conference on Industrial Engineering and Engineering Management (IEEM'09), Dec. 2009, pp. 1588–1592.

Chapter 9

# Applications of TV White Space Networks and Cognitive Radio Techniques to Satellite Communications

Marko Höyhtyä and Miia Mustonen
*VTT Technical Research Centre of Finland Ltd, Oulu, Finland*

## 9.1 INTRODUCTION

Regulatory decisions and spectrum sharing studies conducted in satellite bands in recent years have clearly shown the importance of developing sharing techniques according to satellite system specific characteristics such as long links and transmission latencies, and wide coverage areas. For example, the C band from 3.6 GHz to 4.2 GHz is shared between International Mobile Telecommunications (IMT) and incumbent services, e.g. Fixed Satellite Service (FSS) and Fixed Service (FS). The status of the mobile allocation varies between different International Telecommunication Union Radiocommunication sector (ITU-R) regions and different sub-bands between co-primary and secondary. In Europe (which belongs to ITU-R Region 1), the C band from 3.4 GHz to 3.6 GHz was identified for International mobile telecommunications (IMT) during the World Radiocommunication Conference (WRC) 2012 of the International Telecommunication Union Radiocommunication sector (ITU-R) [1]. Its upper part (3.6 GHz to 4.2 GHz) has also been available on a secondary basis since WRC-12. However, due to primary incumbent usage and its secondary mobile allocation, it has not been possible to use the bands 3.6 GHz to 3.8 GHz exclusively for mobile communications in a harmonised manner in Region 1, especially in Europe. A great deal of spectrum sharing work has been conducted in the terrestrial domain, but due to many differences some of the techniques are not completely applicable in satellite bands, and other techniques need modifications. Some of the most promising spectrum sharing techniques include power control, beamforming and spectrum databases [2–6]. In both terrestrial and satellite domains database techniques have been favoured over spectrum sensing to ob-

**TV White Space Communications and Networks. http://dx.doi.org/10.1016/B978-0-08-100611-5.00012-6**

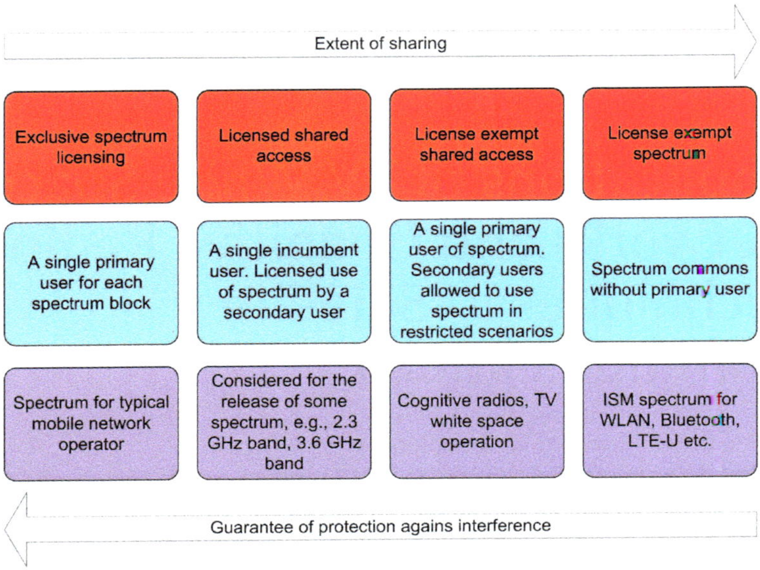


**FIGURE 9.1** Different approaches to spectrum use.

tain awareness of the current spectrum use since they provide better protection to incumbent users.

Conventional satellite operators have several reservations about accepting spectrum sharing in their bands due to competition from new players, potential interference, and limitations on their future technologies and spectrum use, for example. The possible approaches for use of spectrum are shown in Fig. 9.1. Recently, spectrum sharing considerations have evolved from a licence-exempt approach with uncontrolled interference environments towards more controlled set-ups with better operational conditions. To this end, the new Licensed Shared Access (LSA) concept has emerged in European regulation to introduce additional licensed users on bands with existing incumbent licensed users while providing guaranteed quality of service (QoS) for both [7–9]. LSA is a database-based concept that allows sharing in time, frequency and space domains.

TV White Space (WS) spectrum sharing [10–12], which has been described in previous chapters, is also a database-driven approach. As of December 2015, four database operators had qualified to provide database services in the UK where the TVWS concept is perhaps the most advanced globally. Operation in the TV band is hierarchical: the top tier is spectrum licensed to digital terrestrial television (DTT) operators; tier 2 is spectrum used by Programme Making and Special Events (PMSE) devices in locations where frequencies are not used by

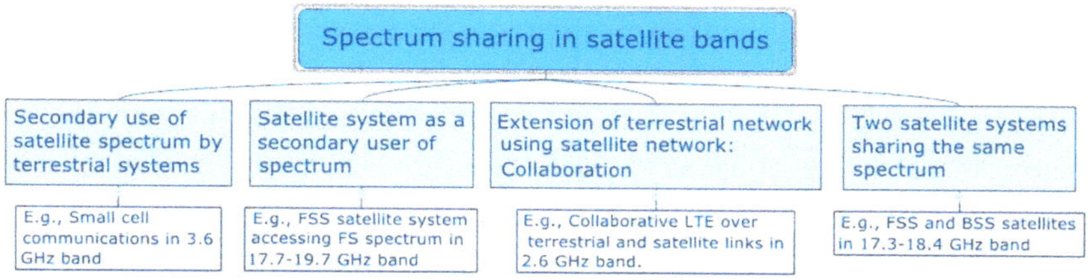


**FIGURE 9.2** Spectrum sharing scenarios in satellite communications.

DTT; and tier 3 is spectrum where White Space Devices (WSDs) are allowed on a licence-exempt basis.

This chapter shows that the database concept can provide a promising method for satellite systems to share the spectrum in controlled manner and to protect their incumbent status in the bands. Although LSA does not provide terrestrial operators the same rights as an exclusive licence obtained via re-farming would, it provides the means to access bands that would not be otherwise accessible in a timely manner. Therefore, LSA is a natural option to reach capacity targets in upcoming 5G environments from both the satellite and terrestrial perspective.

The chapter is organised as follows. In Section 9.2 we introduce cognitive radio techniques applicable in satellite communications. In Section 9.3 we discuss recent regulatory approaches, focusing on licensed sharing. In Section 9.4 we describe the LSA concept for satellite bands, and provides high-level architecture. In Section 9.5 we focus on business aspects. Finally, in Section 9.6 we present the conclusions.

## 9.2 COGNITIVE RADIO TECHNIQUES IN SATELLITE COMMUNICATIONS

This section briefly introduces possible spectrum sharing scenarios in satellite communications and discusses the selected cognitive radio techniques that can be applied in these scenarios.

### 9.2.1 Spectrum Sharing Scenarios

Application scenarios in satellite communications can be classified into four main scenarios (see Fig. 9.2). For each scenario, an example use case and related frequency band is given. In the first two scenarios the sharing follows a traditional cognitive radio approach by using primary and secondary users of the spectrum. The Primary User (PU) is the incumbent of the spectrum, having higher priority or legacy rights on the use of a specific part of the spectrum.

A Secondary User (SU) has a lower priority and therefore exploits the spectrum in such a way that it will not cause interference to the PU. A third scenario defines a case where spectrum sharing is coordinated among satellite and terrestrial users with equal rights. Finally, sharing parties can also be two-satellite systems. In this chapter the focus is on the first scenario.

### 9.2.2 Cognitive Radio Techniques

#### 9.2.2.1 *Spectrum Sensing*

Spectrum sensing can be defined as the task of obtaining awareness about the spectrum utilisation and the existence of PUs in a given geographical area. Frequency, time and spatial domains are typically considered, but also polarisation and directional information can be included to provide more detailed information about spectrum opportunities. Due to the low computational and implementing complexities, energy detection is the most commonly used approach. This approach does not need any prior information of the PU signal and the signal is detected through merely comparing the received signal power with a pre-defined threshold. More advanced approaches include feature detection and matched filter detection, which are more complex to implement but have better sensing accuracy especially with low signal-to-noise ratio values. Feature detection exploits the periodicity or the statistical characteristics of the primary signal, and thus can be used to distinguish different types of PUs [13].

The advantage of sensing is that it supports autonomous operation of SUs. There is no need for data transfer between PUs and SUs. However, a clear disadvantage is that it cannot guarantee interference-free operation. Furthermore, in satellite signal detection, specific infrastructure is required just for sensing since directional high-gain antennas are needed towards the satellite to be able to detect the low power signal [14]. Therefore, sensing is more viable for short-range SUs with limited transmission power.

#### 9.2.2.2 *Databases*

Due to uncertainties and restrictions in sensing, spectrum databases can be seen as a favourable option in many coexistence scenarios. The database may include information about device locations and their activities, spectrum use, coverage/interference levels, relevant regulations and policies, services and networks, etc. This information can be used for several environment-aware network operations such as planning, troubleshooting and radio resource management. Databases allow controlled spectrum sharing where the systems are provided with a predictable QoS. This helps in avoiding unstable situations such as unnecessary frequent channel switching and makes implementation of practical systems easier.

In the first years of research and development for TVWS, spectrum sensing methods were developed and implemented for TV signal detection, and several companies sent their prototype devices to be tested by regulators. After regulators determined that the use of spectrum sensing as a stand-alone method could not sufficiently protect incumbent TV spectrum users, research into the use of spectrum databases began to develop. Currently, databases are the dominant dynamic spectrum sharing approach for TV white spaces. However, there are still several challenges left before databases can be applied to satellite bands since the characteristics of satellite systems and signals differ significantly from terrestrial systems, e.g. regarding received signal power levels and coverage areas.

Recently, there has been active work in the development of databases especially in Ka band (20/30 GHz band), see e.g. [2,15,16]. The database takes information from the national FS registry and also gives information about the FSS terminals, e.g. elevation angle, antenna size and transmission power and calculates where and how the FSS users are allowed to operate in the band. Detailed operational information is found from the given references. In this chapter, we take a detailed look into database-based operation in the 3.6–3.8 GHz band.

#### 9.2.2.3 Beamforming and Smart Antennas

Smart antennas and beamforming techniques enable multiple users to exploit the same frequency resources at the same time and in the same geographical area. With the most advanced active antennas, multiple antenna beams can be provided and controlled independently, improving the spectral efficiency and the system capacity within a given area. While considering the coexistence of a satellite network with the terrestrial cellular network, the reception range of all the satellite terminals is concentrated in an angular sector which can be taken into account when steering the transmission. Beamforming has been recently studied in satellite scenarios (e.g. [17] and [18]). The advantage of beamforming and the use of smart antennas is that this technique enables denser networks and produces less interference in unwanted directions. A disadvantage is the need for more complex and more expensive equipment, and may also require location information from satellite terminals.

#### 9.2.2.4 Beam Hopping

Two multibeam satellites over a common coverage area are studied in [19] where a primary satellite produces larger beams while a secondary satellite has smaller beams. A cognitive beam hopping satellite can enhance the spectrum use assuming that the secondary gateway is aware of the primary's beam hopping pattern. The main difference between conventional and cognitive multibeam systems is that in the latter multiple beams within a cluster share available spectrum in the time domain instead of the frequency domain. Since the primary satellite only illuminates a small fraction of beams out of a large number of beams deployed under beam hopping systems, the rest of the beams remain idle at that time waiting for their transmission slots. Then, another system with smaller beams can operate in the same area and use the free resources. This might only be possible if both satellites are operated by the same operator since the primary operator cannot share the beam hopping sequence with others.

#### 9.2.2.5 *Frequency and Power Allocations*

The main challenge in cognitive radio operation is to take into account all the available information—such as locations of devices, sensing information, regulations, database information etc.—and to make decisions about where in the spectrum to operate at any given moment and how much power to use in that band. Frequency and power should be allocated in a way that optimises the use of available resources while keeping the interference at an acceptable level. In this context, secondary users can adaptively optimise their spectrum use among available channels. A possible disadvantage of adaptive allocation strategies is aggregated interference towards sensitive satellite receivers. Chaotic situations are possible if the secondary systems are autonomously learning their allocations. Thus, care should be taken when developing methods. The transmission power of interfering transmitters has to be limited based on the requirements set in the ITU-R recommendations in order to satisfy interference criteria [16]. In [3] power control of cognitive satellite terrestrial systems is considered. The satellite system has the role of the cognitive system, while the terrestrial represents the primary system.

## 9.3 RECENT REGULATORY APPROACHES

Frequency bands are mainly divided into licensed and licence-exempt bands. While licence-exempt bands are accessible to users quite openly, assuming they will respect transmission power limits and operate according to predefined sharing rules, the licensed bands can only be accessed by users who have obtained a long-term licence to operate in that band. Licence-exempt usage fosters innovations, due to easy access to the spectrum. Licensing on the one hand brings guaranteed QoS, but on the other hand high licence costs impose high market entry barriers to operators. Recently, it has become evident that some parts of the spectrum need to be shared in order to find enough resources for all services. Not all bands can be used exclusively by a single service, but the sharing should preferably be done so that a certain QoS can be guaranteed for all services sharing the band. Different ways to use spectrum are introduced in Fig. 9.1.

Dynamic spectrum allocation using databases is a new and flexible spectrum access method and technology that is used in implementing licence-exempt shared access and in licensed shared access to the spectrum. In TVWS, a WSD will need to contact a White Space Database (WSDB), which will respond to the WSD with a set of operational parameters including the frequencies and maximum power at which the WSD can transmit. In addition, the white space operator will only be permitted to use the white space data to calculate operational parameters if the data are valid. In the United Kingdom, white space data is valid if it constitutes the most recent version made available by the Office of Communications (Ofcom). For example, it is defined in [20] that in the calculation of operational parameters, a WSDB has to take into account updates within four minutes after Ofcom has made them available to the PMSE assignments dataset.

Proposals to address sharing problems in bands other than TV bands have been made in Europe and in the United States, where the latter has focused on the development of a Spectrum Access System (SAS). The main use case for SAS is the Citizens Broadband Radio Service (CBRS) network, which is operated in the 3.55–3.7 GHz band. In this three-tier access model, there are three priority levels for the spectrum use and a database is used to enable sharing between them. Tier 1 users are the incumbent users, tier 2 users operate on a licensed basis and tier 3 users use the spectrum opportunistically. In the case of SAS, the incumbent users are not required to provide any information to a central database, and thus the spectrum awareness information are mainly gathered via sensing [8]. The main incumbents are Department of Defense (DoD) shipborne radar installations operating in coastal areas. FSS systems also operate on this band. SAS maintains a list of locations and look angles of earth station receivers in order to protect their operation.

In Europe the LSA concept is actively advanced by regulatory and standardisation activities. In contrast to the SAS model, the LSA model does not include tier 3 users that are typical in industrial, scientific and medical (ISM) bands such as Wi-Fi systems. LSA is a general regulatory framework that enables shared access to spectrum for a limited number of licensed users. The regulatory framework for LSA to enable cellular mobile network operation in the 2.3–2.4 GHz band has been finalised [21] and standardisation activities are ongoing.

LSA could be applied to allow mobile access to any IMT band in the future in cases where exclusive operation is not possible. An example of such a band in Europe is the 3.6–3.8 GHz band where the incumbent service is FSS. The regulatory work on defining the sharing framework for LSA on this band is ongoing in European regulation. Terrestrial operators consider high-capacity small-cell deployments to be a potential solution for future data demands in this band. Due to WRC-15 decisions, spectrum sharing in the 3.6–3.8 GHz band will be an important study item in the following years. Applying LSA to the band would allow satellite operators to keep their incumbent status in this band. Previously, the 3.4–3.6 GHz band had the same status and satellite operators were against the idea of sharing this band with IMT systems and wanted to have it purely for satellite operations. However, IMT operation was seen to provide more socio-economic benefits and decisions were made to designate this band for mobile services.

Another band which is of high interest regarding database-based spectrum sharing is the Ka band. In 17.7–19.7 GHz, FSS uncoordinated earth stations may operate within CEPT countries on an unprotected basis with respect to the FS activities [15]. FS systems deploy highly directional microwave links and by proper placement of FSS receivers the shared operation in the band is possible. In this band the proposed spectrum sharing approach resembles the TVWS concept [16]. Furthermore, the 17.3–18.4 GHz band, which is designated for feeder links of BSS, can be shared with high-density FSS (HDFSS) systems assuming that BSS information is available in the spectrum database. Finally, after WRC-15 higher frequency bands such as the 37–40 GHz band are also under discussion.

## 9.4 LSA FOR SATELLITE OPERATORS

### 9.4.1 LSA System Architecture

Fig. 9.3 illustrates a possible LSA architecture assuming the satellite operator to be the incumbent of the spectrum. The system relies on a database known as the **LSA repository,** which is usually a centralised database even though research and development for distributed LSA is also ongoing. FSS operators are required to provide a priori information about their spectrum use over the area of interest to this database. Database information includes the radio characteristics of the incumbent user, available bands for LSA operation and band specific transmission power limits, e.g. allowing only small cells to operate in the shared band. The LSA repository is located outside of a specific mobile network.

The **LSA controller** uses this information to either grant access to certain frequency bands or to request band evacuation when the incumbent starts using parts of the band that are currently used by a mobile network. In addition to network external information from the LSA repository, the LSA controller needs network internal information from the operational administration and management (OA&M). This information includes user characteristics and their priorities, network layout, and cell information such as transmission power, locations and antenna patterns [9]. The LSA controller and the OA&M are both part of the mobile network accessing the band as a secondary licensed user. It

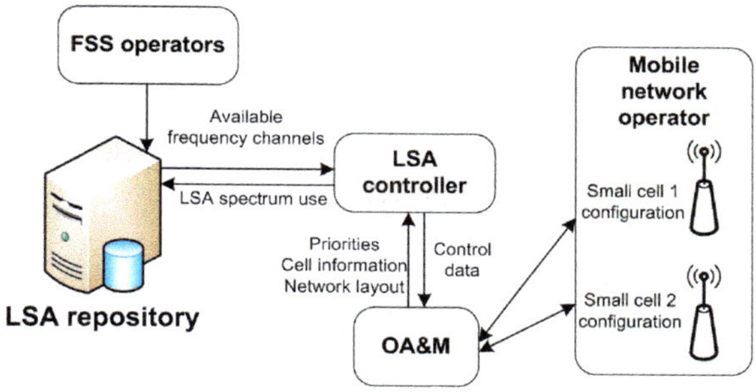


**FIGURE 9.3** LSA architecture for a satellite-terrestrial scenario.

should be noted that since the FSS usage is relatively stable, there might not be a need for specific tools to support mobile access to the band.

### 9.4.2 Important Design Criteria

The main goal of the LSA controller is to use LSA resources as efficiently as possible while avoiding harmful interference to the incumbent. The operations have been quite well defined for terrestrial systems especially in the 2.3–2.4 GHz band (see e.g. [7] and [9]). However, there are new challenges and differences when the concept is adopted in satellite communications. The differences include the large Earth-space distances, which means that the received signal levels are typically low compared to terrestrial signals and require the use of highly directional antennas. In addition, transmission latencies are much longer due to long links which limit the ability to use very dynamic spectrum management approaches. The coverage of a satellite beam is typically orders of magnitude larger than a terrestrial cell, and thus the number of secondary users inside a beam can easily be much higher than in the terrestrial sharing scenario. Finally, the space segment usually needs to be designated once for its entire lifetime. In a GEO system that easily means 15 years of operation with limited possibilities of evolution and maintenance after its launch. Taking these characteristics and limitations into account and following well-defined procedures from terrestrial domain [9], we can outline some important design criteria.

The first phase of the LSA is to negotiate the **sharing framework and the LSA** between the incumbent satellite operator, administration and LSA licensees. The information such as spectrum bands, geographical areas and transmission power limits defined at this stage will remain stable throughout the duration of the LSA agreement. The satellite operator can block certain areas outside the sharing agreement if it seems highly likely that the availability of the LSA band to other services is very low and sharing might be too risky or challenging.

**Network planning** is needed to identify and configure the mobile network cells that will be used during the LSA. This might include definition of the cells inside a certain spot beam of a satellite or smaller area inside a certain beam. At this phase, evacuation times and interference levels with the used antenna patterns can be tested and validated in order to define the LSA controller procedures to be used during actual operation. In addition, the protection distances can be set to allow a specific maximum number of secondary terrestrial transmitters around the satellite receivers.

**LSA resource management** including activation and de-activation of the mobile network cells is needed in a satellite LSA scenario. In contrast to terrestrial networks the satellite operation, especially in downlink, is not as dynamic. However, constant monitoring of the LSA resource availability and interference is needed in order to define which geographical areas need to be evacuated and how the LSA resource use can be optimised. Offloading, frequency handover and other sophisticated techniques can be used to make sure all LSA users will maintain a high QoS. The ability to control and limit the number of users is naturally included in the LSA model since only licensed secondary users are allowed to use the spectrum. A limited number of users obtain the right to use the band while the LSA controller ensures predictable QoS for all spectrum rights holders. This control is important, especially if sharing is considered in the uplink band where a large number of simultaneous users inside the satellite beam can easily raise the interference at the space receiver to unacceptable levels.

### 9.4.3 Example Use Case: C Band (3.6–3.8 GHz)

The previous discussion generally applies to LSA use in the C band. An especially important part of the C band to study is 3.6–3.8 GHz since this has already been allocated to mobile use on a secondary basis in Europe in WRC-12 but due to existing FSS and FS usage it cannot be used in a harmonised manner in Europe. Currently, this part of the band is actively used by FSS systems. Electronic Communications Committee (ECC) project team 1 (PT1) is working to draft an ECC Report providing operational guidelines for spectrum sharing in 3.6–3.8 GHz and, where appropriate, the implementation of LSA at a national level. Depending on the national conditions and the existing usage of the band, the implementation may range from the use of simple protection zones to the application of frequency-specific and location-specific restrictions on the maximum permitted radiated power of individual mobile base station sectors. The study results in [23] suggest that LSA plays a key role in facilitating sharing arrangements between FSS operators and the IMT.

One example of a possible system model is shown in Fig. 9.4, where the incumbent user is the FSS satellite operator and the secondary licensed user is the

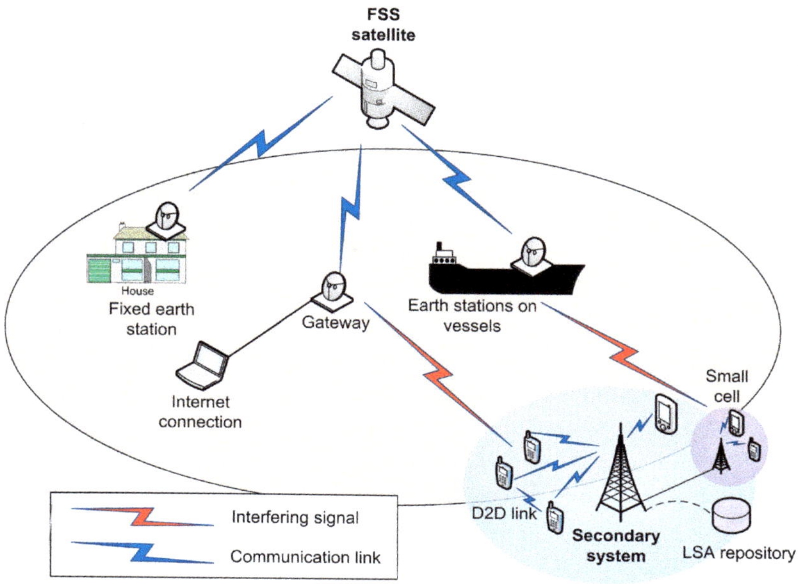


**FIGURE 9.4** C band sharing with LSA.

mobile network operator that is accessing the band using low-power communication, i.e. small cells and device-to-device (D2D) mode. Unlike macro base stations that can require tens of kilometres of protection distances around satellite earth stations the proposed low-power transmissions can operate even closer than 1 km from the FSS very small aperture terminal (VSAT) receivers [22].

Locations of the victim earth stations need to be known in order to enable spectrum sharing using LSA in the C band. This requirement leads to the use of licensed C-band stations in order to know the locations of primary devices. Location awareness technique such as the global positioning system (GPS) is required for SUs as well. Satellite terminals can be either fixed or mobile, as is the case in terrestrial LSA systems. The spectrum database should be kept up to date by the incumbents and by calculating the required interference information based on the given data. Then, both primary satellite users and secondary terrestrial users can experience predicted QoS in an interference-free environment.

Use of LSA assumes that the FSS operator will provide enough information to be able to decide where, when and which parts of the C band are available for secondary use. This may require implementation of an LSA repository and controller that are both included in the ground segment, most probably as a third-party trusted entity. Inclusion of spectrum databases in satellite bands is challenging, and even though the LSA concept seems promising, it needs to be defined carefully from a technical and an economic perspective to determine its real potential. In addition, implementations and field trials should be done, for example to see what the restrictions in time domain are, what the message protocols are, and what information in real implementations might limit the use of this approach. Message protocol development should include all the required system and network elements.

## 9.5 BUSINESS ASPECTS

Opening the C band for sharing may pose some risks to the conventional satellite operators such as competition and interference. Regardless of this, the incumbents may want to continue with their existing business models [24]. On the other hand, high media disruption is likely to lead to increasing service demand, which results in increased spectrum demand and leaves less frequency resources for sharing. This is true for satellite and for terrestrial networks. The role of sharing among incumbents may be emphasised over sharing with mobile operators if the amount of spectrum available for sharing is low.

Spectrum sharing, however, has become more important as a political issue rather than a technical or economical issue as both the European Union and the United States are pushing it forward by high-level decisions. Thus, there is a significant risk in trying to defend the current assets and position without considering sharing. In the worst case the political pressure might eventually lead to losing the spectrum assets to other wireless services with higher societal value. Controlled sharing is thus an attractive option to save the use of spectrum for the current services and to save the rights of use as an incumbent operator. As is discussed in [24], sharing based on LSA is an opportunity for incumbents to show their cooperative attitude and societal responsibility by improving the efficiency of spectrum use. By allowing sharing, incumbents can continue their operations in the bands to fulfil their societal obligations defined with minimal additional investment.

An important part of the inclusion of flexible spectrum sharing for incumbent operators is keeping the intelligence in the network in order to be in a controlling position. However, in the FSS case the environment is not that dynamic, and quite simple procedures can be used. An interesting analysis of business opportunities in spectrum sharing for operators including the definition of four different business models is given in [25]. The paper defines rules for incumbent operators that we have modified, in part to be more appropriate for the already used satellite bands. The modified rules are given in Table 9.1. The main parts of the table consider both terrestrial and satellite incumbent operators.

**TABLE 9.1** Spectrum Sharing Rules for an Incumbent Operator, Based on [25]

| Opportunity | How-to | Boundary | Priority | Timing / Strategy |
|---|---|---|---|---|
| –Utilise existing customer base and growing demand | –Invest to maintain customer lock-in and dominant market position | –Leverage installed infrastructure and standard evolutions | –Retain control over spectrum and cognitive radio techniques in the network | –Optimise own spectrum use first |
| –Get faster access to new lower-cost spectrum | –Buy all exclusive spectrum available | –Maintain direct contacts to regulators | –Start sharing from other bands than own if possible | –Then consider sharing with others |
| –Utilise local areas with enhanced small cells for QoS and offloading | –Combine spectrum assets to deliver higher data rates | –Turn entrant operators to virtual ones (e.g. in satellite-satellite sharing scenario) | –Appreciate revenue over efficiency | –Do not give up the whole spectrum, rather share before losing |

The main opportunity in the business models based on spectrum sharing for current incumbent operators lies in the exploitation of the existing customer base and their growing data and service demands. There is a possibility to rapidly access new lower-cost spectrum in higher bands, enabling the utilisation of the dominant position in the market. Thus, investments should aim to strengthen the customer lock-in and the market position. Direct connection and discussion with the regulator is very valuable in order to remind the regulator that earlier investments in the networks have pay periods that are sufficiently long. It is essential for satellite operators to have long-term access to the spectrum to gain enough revenue from the operations.

The preferred case would be to have the regulator actively promoting sharing and defining the value of the shared spectrum with the aim of improving the efficiency of spectrum use. Taking into account the given discussion and all the aspects described in the table, the following business aspects need to be covered in the incumbent operator strategies:

1. **Service creation and value proposition**: The service offered to the end users should be new or better than what is available in the market. An incumbent operator should be able to provide improved or new services to the users to strengthen its position in the frequency bands. The offering should create value that can be either quantitative (e.g. price, time) or qualitative (e.g. customer experience).
2. **Define customers and the revenue streams**: The types of customers for which the service offer is useful or of special interest should be identified. The incumbent operator should define how it earns money through the operations. For example, an operator obtaining spectrum fees from sharing could provide data with lower price to the end users and still receive higher total revenue than in a case where it is the only operator in the spectrum.
3. **Cost structure**: An incumbent operator should estimate the components in the cost structure in terms of investments and running costs related to networks, services, customers, and marketing, for example how much the transponders covering new frequency bands would cost or what the price of LSA repository and controller would be and how the costs should be shared among sharing operators. In general, the incumbent does not pay for these services; the implementation of the sharing is the responsibility of the entrant system (mobile operator) and possible national regulatory authority.

In contrast, the most likely opportunity for the entrant systems would be in the utilisation of new, specialised demand that is not offered by the incumbent operators. Complementary services may open up a clear role for the entrant systems. Here, it could be providing terrestrial Internet access in the same band.

## 9.6 CONCLUSIONS

In this chapter, we discuss the application of TV white space communications and cognitive radio techniques to satellite communications, focusing on the use of spectrum databases. One of these approaches is the LSA concept, which also seems promising for spectrum sharing in the satellite bands, which is under discussion in European regulation. LSA provides a framework and methods that enable the sharing parties to experience predictable QoS while ensuring higher usage rights to the current incumbent users. This includes the right to request LSA users (such as a mobile operator) to terminate transmission in the shared band at any time and in any geographical area. However, taking into account the stable nature of satellite usage, it is not foreseen that the evacuation of the band will be a major issue. In this chapter, we introduced a potential LSA architecture, defined some critical aspects of functionality and discussed the business aspects of the concept in a satellite band. The C band was given as a potential use case for the concept, and the work going on in this band in Europe and the United States was introduced. The chapter gives an overview of the concept; however, more technical work and economic studies are needed to reveal its full potential of the LSA concept for all involved parties.

## ACKNOWLEDGEMENT

This work was performed in the framework of the CORE++ and CORNET projects, partially funded by Tekes, the Finnish Funding Agency for Innovation. In addition, part of the work was carried out under the FREESTONE project funded by the European Space Agency.

## REFERENCES

[1] ITU Radio Regulations, edition of 2015.

[2] S. Maleki, S. Chatzinotas, B. Evans, K. Liolis, J. Grotz, A. Vanelli-Coralli, N. Chuberre, Cognitive spectrum utilization in Ka band multibeam satellite communications, IEEE Communications Magazine 53 (March 2015) 24–29.

[3] S. Vassaki, M.I. Poulakis, A.D. Panagopoulos, Optimal iSINR-based power control for cognitive satellite terrestrial networks, Transactions of Emerging Telecommunications Technologies (2015).

[4] M. Höyhtyä, J. Kyröläinen, A. Hulkkonen, J. Ylitalo, A. Roivainen, Application of cognitive radio techniques to satellite communication, in: Proc. DySPAN, October 2012, pp. 540–551.

[5] S.K. Sharma, S. Chatzinotas, B. Ottersten, Satellite cognitive communications: interference modelling and techniques selection, in: Proc. ASMS, September 2012, pp. 111–118.

[6] M. Höyhtyä, J. Ylitalo, X. Chen, A. Mämmelä, Use of databases for dynamic spectrum management in cognitive satellite systems, in: Cooperative and Cognitive Satellite Systems, Wiley, 2015.

[7] ECC Report 205, Licensed shared access, February 2014.

[8] M.D. Mueck, S. Srikateswara, B. Badic, Spectrum sharing: licensed shared access (LSA) and Spectrum Access System (SAS), White paper, 2015.

[9] M. Mustonen, M. Matinmikko, M. Palola, S. Yrjölä, K. Horneman, An evolution toward cognitive cellular systems: licensed shared access for network optimization, IEEE Communications Magazine 53 (May 2015) 68–74.
[10] Ofcom, Implementing TV white spaces, Statement, February 2015.
[11] A.B. Flores, R.E. Guerra, E.W. Knightly, P. Ecclesine, S. Pandey, IEEE 802.11af: a standard for TV white space spectrum sharing, IEEE Communications Magazine 51 (October 2013) 92–100.
[12] G. Ding, J. Wang, Q. Wu, Y.-U. Yao, F. Song, T.A. Tsiftsis, Cellular-base-station-assisted device-to-device communications in TV white space, IEEE Journal on Selected Areas in Communications 34 (January 2016) 107–121.
[13] F. Dimc, G. Baldini, S. Kandeepan, Experimental detection of mobile satellite transmissions with cyclostationary features, International Journal of Satellite Communications 33 (Mar.–Apr. 2015) 163–183.
[14] M. Höyhtyä, Secondary terrestrial use of broadcasting satellite services below 3 GHz, International Journal on Wireless & Mobile Networks 5 (February 2013) 1–14.
[15] ECC Report 232, Compatibility between Fixed Satellite Service uncoordinated receive Earth Stations and the Fixed Service in the band 17.7–19.7 GHz, May 2015.
[16] M. Höyhtyä, A. Mämmelä, Spectrum database for coexistence of terrestrial FS and FSS satellite system in the 17.7–19.7 GHz band, in: Proc. KaConf, Oct. 2015.
[17] C. Yuan, M. Lin, J. Ouyang, Y. Bu, Beamforming schemes for hybrid satellite-terrestrial cooperative networks, International Journal of Electronics and Communications 69 (August 2015) 1118–1125.
[18] S.K. Sharma, S. Chatzinotas, B. Ottersten, Transmit beamforming for spectral coexistence of satellite and terrestrial networks, in: Proc. CrownCom, July 2013, pp. 275–281.
[19] S.K. Sharma, S. Chatzinotas, B. Ottersten, Cognitive beam hopping for spectral coexistence of multibeam satellites, International Journal of Satellite Communications and Networking 33 (January/February 2015) 69–91.
[20] Ofcom, White Space Database Provider (WSDB) Contract, Draft, 2015.
[21] CEPT Report 58, Technical sharing solutions for the shared use of the 2300–2400 MHz band for WBB and PMSE, July 2015.
[22] M. Höyhtyä, Sharing FSS satellite C band with secondary small cells and D2D communications, in: Proc. ICC Workshops, June 2015, pp. 9666–9671.
[23] Plum consulting, Use of C-Band (3400/3600–4200 MHz) for mobile broadband in Hungary, Italy, Sweden and the UK, June 2015.
[24] P. Ahokangas, M. Matinmikko, S. Yrjölä, M. Mustonen, E. Luttinen, A. Kivimäki, J. Kemppainen, Business scenarios for incumbent spectrum users in licensed shared access (LSA), in: Proceedings of the CrownCom, Oulu, Finland, 2014, pp. 407–412.
[25] P. Ahokangas, M. Matinmikko, S. Yrjölä, H. Okkonen, T. Casey, Simple rules for mobile network operators' strategic choices in future cognitive spectrum sharing networks, IEEE Wireless Communications 20 (April 2013) 20–26.

# Index

CPI Antony Rowe
Chippenham, UK
2018-02-16 10:26